Springer-Verlag Berlin Heidelberg GmbH

Christian Schwabe   Erika E. Büllesbach

Relaxin and the Fine Structure of Proteins

Springer

Christian Schwabe

Medical University
of South Carolina
Charleston, South Carolina,
U.S.A.

Erika E. Büllesbach

Medical University
of South Carolina
Charleston, South Carolina,
U.S.A.

ISBN 978-3-662-12911-1

Library of Congress Cataloging-in-Publication data

Schwabe, Christian.
    Relaxin and the Fine Structure of Proteins / Christian Schwabe, Erika E. Büllesbach
        p. cm. — (Biotechnology intelligence unit)
    Includes bibliographical references and index.
    ISBN 978-3-662-12911-1   ISBN 978-3-662-12909-8 (eBook)
    DOI 10.1007/978-3-662-12909-8

    1. Relaxin.   I. Büllesbach, Erika E., 1948-. II. Title. III. Series
        [DNLM: 1. Relaxin—chemistry. 2. Relaxin—metabolism. WP 530 S398r 1998]
    QP572.R46S39   1998
    573.6'654—dc21
    DNLM/DLC   97-52386
    for Library of Congress   CIP

Typesetting: R.G. Landes Company, Georgetown, TX, U.S.A.

SPIN 10681638   5 4 3 2 1 0 - Printed on acid-free paper

ACKNOWLEDGMENTS

Contributors to this book are the members of our small, efficient and enthusiastic research team including Barbara Rembiesa, Robert Bracey, and George Fullbright. Jutta Schwabe was the relentless editorial whip and Rosemary Taylor decoded dictations and translated them into understandable language. Bernard G. Steinetz has read the manuscript and lifted our spirit with well-timed, eloquent flattery. Drs. Roy O. Greep and Robert Krock continuously fanned that little flame that burns at the beginning of every project and have instilled in us the feeling of importance of our relaxin-related efforts. These researchers, of course, hark back to the laboratory of the discoverer of relaxin, F. L. Hisaw, and thus we are grateful to him, whom we never knew, for educating the significant researchers in this field and for giving us relaxin.

The Institute of General Medical Sciences of NIH has supported these authors' experimental work when the potential outcome was more in the balance than it is today, and the Connetics Biotechnology Company has contributed significantly to the success of this work. Rough spots in the authors' careers have been smoothed over by generous support from the MUSC University Research Committee.

Inasmuch as all acknowledgments seem to contain tales of a deed without which a particular book could not have been written, we must add the editors of Landes-Springer Verlag who invited this relaxin book—a project that would not have even entered our minds without their friendly nudge.

Many thanks to all of you.

FOREWORD

Whoa. I sense the presence of Relaxin, a substance that has been swept under the rug for lo these many years solely on the basis of an undeserved prejudicial attitude on the part of the endocrine community. Relaxin was somehow held not to be a fully sanctified hormone and thus not fully qualified for membership in the endocrine society. Here the entire story of relaxin has been told in a style which is not only bold and clarion but also lively and at times even whimsical. In addition, the whole pie has been sprinkled with quotes from the classic literature. So this book is a wonderful blend of the best of science and the authors' philosophical attitude on science in general and the conduct of research in particular. In all it makes for fascinating reading.

Another helpful feature of this book is the excellent organization of very complex material. To wit: chapter 8, "Total Synthesis of Human Relaxin", provides a close look at the method used to achieve the purity of the final product. This is a highly technical procedure. What the authors have done is set aside the next several chapters to deal individually with the role of some of the most important amino acids: alanine, glycine, arginine, citrulline, and methionine. Thus, they averted what could have been a case of biochemical chaos. I have only occasionally played an active role in the development of research on relaxin. During my first days with Dr. Frederick Hisaw, the discoverer of relaxin, I measured, as best one can, the degree of separation of the pubic symphysis in rodents treated with relaxin. Later, a Harvard medical summer student, Dr. Bernard Kliman, came to work in my laboratory, and mainly through his effort we managed to develop the mouse bioassay for relaxin.

I have also had contact with research on relaxin through manuscripts and reprints sent to me. Relaxin is species specific, and each species has a relaxin molecule with its own unique amino acid sequence. Now one must speak of rabbit relaxin or human relaxin, as the case may be. Relaxin has been found in an astonishing variety of species, ranging from the invertebrates to common laboratory mammals and some wild animals. A true hormone, relaxin is blood borne; its receptors are widely distributed among different body tissues.

The Darwinian concept of the origin of life is based on a single point hypothesis: all life evolved from a primordial organism and humans evolved from an early African woman. The genomic hypothesis of the author and a few of like mind hold that there are multiple possible points of origin. The authors argue that relaxin is an example of the genomic potential hypothesis. Considering that the Darwinian hypothesis has acquired a strong root system over the past one and a half centuries, the difference of opinion held on the original of all life may well turn out to be the debate of the 21st century. The authors are to be congratulated for helping to set in motion this exciting debate.

Roy O. Greep, Ph.D.
Harvard University

CONTENTS

CHAPTER 1

Introduction

Relaxin, perhaps more than any other hormone, has a varied history of eliciting enthusiasm, rejection, skepticism, and long lapses of neglect. Hisaw discovered the factor in 1925 in the serum of pregnant guinea pigs which, if injected into virgin guinea pigs, caused remodeling and widening of the symphysis pubis. Five years later Fevold et al demonstrated that the new factor was proteinaceous and that it could be isolated from ovaries.

The reader, perhaps students in particular, may be interested to learn how thin and frail the thread may be that ties our efforts to success. Many years after this discovery we learned that the effect of guinea pig relaxin is unusually species-specific. Had Hisaw used mice or rats for the test, the relaxin factor may have remained undiscovered for many more years. It seems peculiar that guinea pig relaxin should be so ineffective in other species in light of the fact that the guinea pig relaxin receptor recognizes almost every other molecule of this genre.

One prejudice of the time was that ovaries contained no peptide hormones. When Fevold actually isolated relaxin from pregnant pig ovaries, a wave of skepticism went through the community of endocrinologists which lasted for many years. As late as 1974 when relaxin was a white, water soluble powder in our laboratory an endocrinologist gently suggested that our research might be wasted on a figment of imagination which might at best be a steroid tightly adsorbed onto a protein. These prejudices were finally laid to rest when the sequence of porcine relaxin was published in 1976-77.

For a long time after its discovery relaxin has been a hormone in search of a physiological role in humans. There was little doubt that relaxin was required for live birth in rats and other rodents as well as in some of our agriculturally important species such as pigs

Relaxin and the Fine Structure of Proteins, by Christian Schwabe and Erika E. Büllesbach. © 1998 Springer-Verlag and R.G. Landes Company.

and cows, but a clear-cut need for relaxin in humans was never established. Instead, the number of possible functions of relaxin increased rapidly with the availability of purified material. Today relaxin is implicated in mammary growth, development and functions, ionotropic and chronotropic action on the heart, and angiogenic activity in the peripheral tissues as well as in the endometrium, but it is best known for its effect on the uterus and the symphysis pubis in preparation for parturition. Suddenly there were too many functions; however, the literature suggests that not all occur in one species, and that observation adds yet another interesting facet to the characteristics of this molecule.

Chemical characterization of relaxin revealed a two-chain molecule with a disulfide bonding pattern identical to that in insulin but with very little primary sequence homology. This observation caught the attention of molecular evolutionists who saw clear evidence for gene duplication as a cause for the existence of relaxin and insulin. From a structural point of view it seemed amazing that relaxin and insulin, with 70% primary sequence difference, could still attain similar secondary structures such as to give rise to nearly identical circular dichroism (CD) spectra. Even mouse relaxin, which sports one extra residue within the large interchain ring, is essentially helical and indistinguishable from other relaxins by this method. The exquisite sensitivity of CD measurements to helix content produce astounding results and regularly elicit a response from those interested in predicting three-dimensional structures from primary sequences.

Apart from the interest in relaxin as a hormone the structure function relationship of this molecule has sent quite a few ripples through the ranks of protein chemists outside the relatively small community of relaxin researchers. Rarely has a protein taught so much about fine structural requirements for its active conformation.

There are several "first" surprises and anti-dogmas in the relaxin story that need be discussed in order to transmit to the reader some of the excitement, and even adventure, that surrounded the development of the relaxin field. This goal may clash with the etiquette of straight-laced science editorial policy but it makes for more interesting reading. Consequently, this will be a story with a message supported by selected experiments rather than a catalogue of names and events.

Less emphasis will be given to physiology because the field is covered quite well by recent (some voluminous) reviews—all of which give a good description of the state of affairs which reflects a profusion of activities of relaxin in different species. Multiple biological effects alluded to earlier may in part be due to the species specificity of the relaxin structures and due to the fact that mainly porcine relaxin was used to perform most of the reported studies, particularly in rodents. Does one have to consider another level of fine-tuning in order to resolve the enigma of multiple, often unrelated, actions of relaxin in various species? Would it be necessary to use rat relaxin when rat physiology is under investigation and mouse relaxin for mice and so on?

That would put quite a burden on researchers because only porcine relaxin can be isolated in sufficient quantities to perform all the studies required to learn about the mode of action of a hormone. This is one of the many reasons for slanting this book in the direction of relaxin chemistry and structure function work not often seen in the literature, although such expertise is required to remedy the specificity problem. We have published detailed methods for the synthesis of these two chain disulfide-linked small proteins and have made some under contract for industry and for researchers who were totally unfamiliar with the techniques of peptide chemistry and needed a specific relaxin for their animal model. Thus, if physiology of relaxin is on one's mind, the native molecule for any species is not sufficiently out of reach to justify work with a non-homologous hormone. Examples are beginning to accumulate that support this experience-inspired conservative view. Porcine relaxin inhibits spontaneous contractions of human uterine muscle in vitro but human relaxin will not. The same holds for the cervical ripening in humans which has puzzled a Darwinistically oriented society for some time.

Many of the effects observed, when porcine relaxin is used in rats, may therefore not be as relevant as once thought. Serendipity provided an even more stringent restriction in case of the rat model. Isolated rat relaxin had a much lower potency than synthetic rat relaxin which ranked high with the human and porcine molecules in the mouse assay. It is not clear precisely if something happened to this molecule during isolation or whether the preparation was simply not pure, but its reported bioactivity is significantly below that of synthetic rat relaxin, human relaxin, and native porcine relaxin in the same assay. Just a note of caution.

The pharmacological action of the hormone is quite a different issue. Here the differences and potency and effectiveness may pay enormous dividends. If prejudice against heterologous therapy can be overcome we should consider ourselves fortunate to have a large arsenal of different relaxin structures available. Salmon calcitonin, for example, is used in humans to treat Padget's disease, but will the U.S. Food and Drug Administration (FDA) permit treatment of humans with rat or porcine relaxin? That may be a tough call, but when presented with a choice of either severely debilitating disease or rat relaxin injections, the rat may suddenly acquire the glow of conditional sainthood.

The prejudice has two reasons, one concerns antibody formation and the second is due to the Darwinian notion that the endogenous molecule is always better than the exogenous one because evolution has put them together via a succession of adaptive mutations.

The concern about antibodies is theoretically correct, but in practice, much less problematic than anticipated. Cloned human insulin is not significantly better than porcine insulin as concerns antibodies but is used in preference to either porcine or bovine insulin because of an advertising campaign that addresses the species issue and is commercially motivated.

All of this boils down to the fact that for physiological studies the endogenous ligand should be used exclusively whereas for therapeutic purposes the most potent molecule is appropriate, regardless of the animal of origin. In this book we describe how to get the necessary molecules or derivatives for research, and that kind of reasoning extends right into the design of new molecules for specific applications. As this story develops new data are coming from the laboratory almost weekly. Hopefully, the reader will get a little taste of the excitement of discovery at the most fundamental level of biology—the catalytic surface of proteins.

CHAPTER 2

A Very Brief History

It stands to reason that a fetus, as it nears completion, needs a passage to the outer world that is commensurate with its size exactly at the time of parturition. The discrepancy between the pelvic canal and the fetus in most mammals is so large and the pelvis so rigid that a special mechanism must exist which corrects this discrepancy if a species is to persist. Here is one of the few examples where biology has to submit to reason. The birth canal cannot be opened during gestation because of the danger of premature fetal loss, and at the end of gestation the opening of the canal must be coordinated with a dramatic over-all shift of the endocrine balance which causes potentially damaging uterine contractions. This set of circumstances is well known to endocrinologists and explains Hisaw's experiments with pregnant guinea pig serum. Only a hormone could cause such a change, and most hormones travel from the gland of origin to the site of action by the blood stream. Serum from pregnant animals with relaxed symphyseal joints was a plausible place to search for this factor. Once virgin guinea pigs had been treated with estrogen they responded to the pregnancy serum by significant widening of the symphysis pubis.[1] This was a momentous discovery albeit overshadowed by the discovery of insulin, the pancreatic disulfide homologue of relaxin, at about the same time. Insulin took center stage at once because it saved lives and relaxin fell by the wayside because humans had developed a surgical procedure to overcome the problems of pelvic insufficiency. Nonetheless, relaxin researchers had an inkling that insulin and relaxin had something in common (Fig. 2.1). A condensed pictorial summary of both factors as they evolved through their stages of discovery is shown in Figure 2.2.

Insulin research made tremendous progress early on and stayed well ahead of our understanding of relaxin structure and function until 1980. Now the lines have merged and although relaxin action

Relaxin and the Fine Structure of Proteins, by Christian Schwabe and Erika E. Büllesbach. © 1998 Springer-Verlag and R.G. Landes Company.

COMPARISON OF THE PROPERTIES OF INSULIN AND RELAXIN		
	Insulin	Relaxin
Isoelectric point	About PH 5.0	PH 5.4–5.5
Acid solution	Stable	Stable
Alkaline solution	Unstable	Unstable
Millon's reaction	Questionable	Negative
Biuret reaction	Positive	Questionable
Molisch Reaction	Negative	Negative
Picric acid	Precipitated by	Precipitated by
Trypsin	Destroyed	Destroyed
Pepsin	Destroyed	Destroyed

Fig. 2.1. Fevold et al compared the properties of relaxin and insulin in 1930. Reprinted with permission from: Fevold HL, Hisaw FL, Meyer RK. J Am Chem Soc 1930; 52:3340-3348. Copyright 1930 The American Chemical Society.

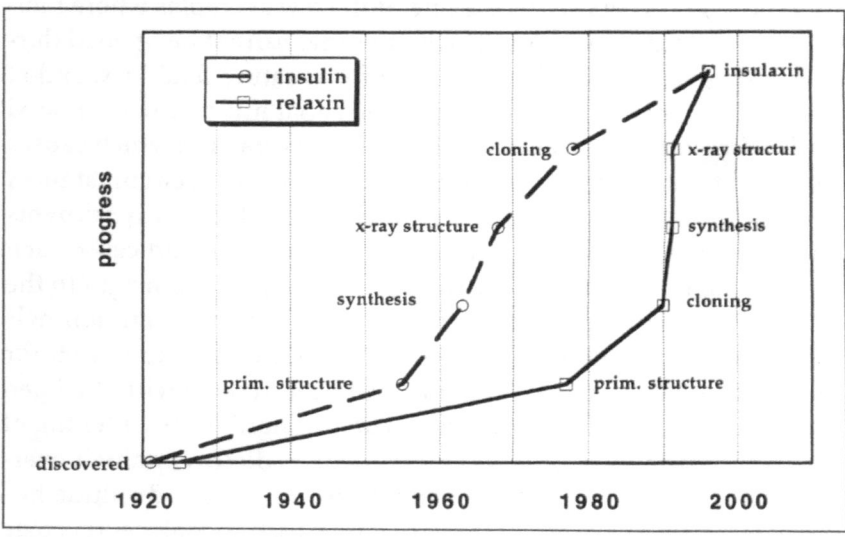

Fig. 2.2. The development of two hormones, insulin and relaxin, starting with their discovery in the 1920s to the Zwitterhormon, insulaxin, which bears both activities. Only the major biochemical events are emphasized that contributed to the understanding of the structure and provide the bulk of hormone needed for pharmaceutical applications.

in humans is still not well defined, we have to date learned more about its structure and function and its active site than is known about insulin.

Not to diminish the many contributions made by whole animal and cell physiologists, the most remarkable event separating the time of experimentation with relaxin-enriched tissue extracts and the modern structure function era was a pharmacological effect of relaxin observed by Casten and Bouzec in the late 1950s to mid 1960s.[2] A crude extract made from pregnant pig ovaries was given to patients with peripheral vascular insufficiencies and to some with scleroderma. One of the consistent observations was the increase in skin temperature and the rapid healing of trophic ulcers. The studies were performed without proper controls and published in a less than conspicuous journal. All of this contributed to the fact that these studies were largely discredited and generally forgotten until, and almost on the heels of a negative outcome of a clinical trial of human relaxin in late pregnancy, a new company was formed to explore the potential of relaxin in scleroderma treatment. The first clinical research will come to an end in summer 1997 so that the reader might look for comments pertaining to the outcome near the end of this book (chapter 18).

The vascular effects of human relaxin appear to be less when compared to the original observation which may be ascribed to the structural differences of the relaxins used in these studies. As alluded to in the introduction, human relaxin may not be the most effective mediator of the vascular effects in humans so that porcine relaxin for example may have to be explored again for that purpose. The importance of following up Casten and Bouzec's lead cannot be overemphasized in light of the fact that diabetes is a rapidly increasing condition in the United States and that every year 50,000 cases of surgical intervention are required due to the peripheral vascular changes associated with the late stages of diabetes. It is a moral obligation of relaxin researchers to provide a definitive answer concerning this aspect of potential relaxin action.

The new era began essentially with the first large-scale purification of relaxin from ovaries of pregnant sows in 1974[3] which culminated in the sequence analysis and disulfide structure identification in 1976[4] and 1977.[5] Since that time the receptor-binding site of relaxin has been delineated with certainty and some of the most important fine structural features that shape different conformations for insulin and relaxin have been discovered.[6,7] This knowledge in turn has led to the synthesis of a bona fide "Zwitterhormon", called insulaxin, which binds competitively to both the relaxin and the insulin receptors in the rat brain and human placentas, respectively.[8]

Relaxins as defined in this book have been isolated from a variety of animals of different families and serological evidence suggests that relaxin exists in different phyla and kingdoms as well, i.e., in invertebrates and possibly protozoans,[9] plants and bacteria (unpublished results). Relaxin and insulin-like two-chain molecules are found in different tissues of many forms of life from insects[10,11] to humans where they act in unrelated types of function. These molecules should be of great interest to students of structure and function relations and to cosmopolitans because they do carry the message that the presently accepted view of evolution is fundamentally flawed.

References

1. Hisaw FL. Experimental relaxation of the pubic ligament of the guinea pig. Proc Soc Exp Biol Med 1926; 23:661-663.
2. Casten GG, Boucek RJ. Use of relaxin in the treatment of scleroderma. J Am Med Assoc 1958; 166:319-324.
3. Sherwood OD, O'Byrne EM. Purification and characterization of porcine relaxin. Arch Biochem Biophys 1974; 160:185-196.
4. Schwabe C, McDonald JK, Steinetz BG. Primary structure of the A chain of porcine relaxin. Biochem Biophys Res Commun 1976; 70:397-405.
5. Schwabe C, McDonald JK, Steinetz BG. Primary structure of the B-chain of porcine relaxin. Biochem Biophys Res Commun 1977; 75:503-510.
6. Büllesbach EE, Schwabe C. On the receptor-binding site of relaxin. Int J Peptide Protein Res 1988; 32:361-367.
7. Büllesbach EE, Yang S, Schwabe C. The receptor-binding site of human relaxin II: A dual prong-binding mechanism. J Biol Chem 1992; 267:22957-22960.
8. Büllesbach EE, Steinetz BG, Schwabe C. Chemical synthesis of a Zwitterhormon, insulaxin, and of a relaxin-like bombyxin derivative. Biochemistry 1996; 35:9754-9760.
9. Schwabe C, LeRoith D, Thompson RP et al. Relaxin extracted from protozoa *(Tetrahymena pyriformis)*. J Biol Chem 1983; 258:2778-2781.
10. Ishizaki H, Suzuki A. An insect brain peptide as a member of the insulin family. Horm Metabol Res 1988; 20:426-429.
11. Geraerts WPM, Smit AB, Li KW et al. The light green cells of lymnaea: a neuroendocrine model system for stimulus-induced expression of multiple peptide genes in a single cell type. Experientia 1992; 48:464-473.

CHAPTER 3

The Relaxin Insulin-Like Motif in Protein Structures

A t least one of the authors is old enough to remember how puzzled protein chemists were to see a distinct insulin structure in lieu of any of 11 other possible cross-link isomers. At that time students would also think about such problems and one of them actually presented a hypothesis concerning this phenomenon plus experiments as a research proposal to his examination committee. This was several years before Steiner and his colleagues discovered a single-chain proinsulin.[1] The student's rationale was that a monomolecular reaction (single chain) is always faster than a bimolecular one and that the folding of the single chain could force the disulfide bonds always into the same parallel direction as they are found in insulin and relaxin (which was not known at that time). This proposal hit the examining committee like Jason's stone hit the dragon man. They were fighting among each other until they realized about an hour later that the candidate was off the hook. Diffusion-controlled folding of a single chain molecule was many orders of magnitude faster (10^{11} per second) than the actual process of protein synthesis which occurs at a rate of approximately one or two residues per minute in mammalian cells and up to 4 residues per second in prokaryotes. In either case a growing chain can sample all possible energy conformations in the time it takes to add another residue and can select momentarily the lowest energy valley at any stage of chain growth. This means that the product may not necessarily be in the lowest cumulative energy valley at the end of the folding process which becomes apparent as the molecule rearranges itself slightly after connecting peptides are split from the chain. The physical chemist on the committee suggested that this idea be published in one of the highly visible journals but the thesis advisor, a great person in every respect,

Relaxin and the Fine Structure of Proteins, by Christian Schwabe and Erika E. Büllesbach. © 1998 Springer-Verlag and R.G. Landes Company.

had grown up with the standards of the Lutherans in Lake Wobegone, i.e., "publishing ideas is immodest; stick to experiments." In retrospect this is a peculiar attitude since science is primarily about ideas.

The ideas turned out to have been on target; the credit will not always come your way as a student or even later in a career, but the knowledge that one can use the mind to conceptualize is there and next time or the time thereafter perhaps it will work.

The discrepancy of pelvic and fetal size indicated a fundamental problem which was relatively easy to recognize and to assess. Surprisingly only Hisaw and his group seized upon this problem. The other surprise was that the mediating hormone relaxin should have been a structural homologue of insulin.

The structural motive of two small peptide chains linked by disulfide bonds has since been observed relatively frequently. The methods of molecular biology allow for quick scanning of the coding potential of genomic DNA and one of the first structures discovered this way was the relaxin-like factor found in the Leydig cells of humans, mouse, pig, cattle and sheep. The name Leydig insulin-like factor[2,3] changed to relaxin-like factor (RLF) once the molecule had been synthesized and the protein revealed greater affinity to relaxin.[4] A different factor called INSL-4 was discovered by screening a cDNA library of human placenta[5] but in this case subsequent synthesis exposed, if anything, a more insulin-like character.

In the silk worm *Bombyx mori*, a relaxin-like disulfide structure appears to be a developmental factor at least when administered to the pupae of a related insect species,[6] but bombyxin receptors found in the ovaries may indicate a more relaxin-like function.[7,8] Mollusks (snails) possess insulin-like proteins[9] and although this factor has a fourth disulfide bond it is structurally not far removed from the relaxin that occurs in mammals. Elasmobranchs have relaxin in the ovaries and possibly in the testes and, surprisingly, this hormone does elicit relaxation in the symphysis pubis of the guinea pig and to a lesser extent in the mouse which is remarkable because the shark does not possess such a structure. The relaxin effect in cartilaginous fishes is not known with certainty particularly not in oviparous species such as the skate *Raja erinacia*. The sandtiger shark *Odontaspis taurus* and the spiny dog fish *Squalus amcanthias* are ovoviviparous and here the exit of live young could be facilitated by relaxin.

A relaxin-like immuno-activity has been observed in single cellular organisms such as tetrahymena[10] and insulin-like activity has even been shown to exist in plants.[11]

It seems intriguing that a structure with a stringent demand upon the precision of disulfide bond synthesis should occur so often in such diversified forms of life.

References

1. Steiner DF, Oyer PE. The biosynthesis of insulin and a probable precursor of insulin by a human islet cell adenoma. Proc Natl Acad Sci USA 1967; 57:473-480.
2. Adham IM, Burkhardt E, Benahmed M et al. Cloning of a cDNA for a novel insulin-like peptide of the testicular Leydig cells. J Biol Chem 1993; 268:26668-26672.
3. Burkhardt E, Adham IM, Brosig B et al. Structural organization of the porcine and human genes coding for a Leydig cell-specific insulin-like peptide (LEY I-L) and chromosomal localization of the human gene (INSL3). Genomics 1994; 20:13-19.
4. Büllesbach EE, Schwabe C. A novel Leydig cell cDNA-derived protein is a relaxin-like factor (RLF). J Biol Chem 1995; 370:16011-16015.
5. Chassin D, Laurent A, Janneau JL et al. Cloning of a new member of the insulin gene superfamily (INSL4) expressed in human placenta. Genomics 1995; 29:465-470.
6. Ishizaki H, Suzuki A. Brain secretory peptides of the silkmoth *Bombyx mori*: Prothoracicotropic hormone and bombyxin. [Review]. Progr Brain Res 1992; 92:1-14.
7. Tanaka M, Kataoka H, Nagata K et al. Morphological changes of Bm-N4 cells induced by bombyxin, an insulin-related peptide of *Bombyx mori*. Regulatory Peptides 1995; 57:311-318.
8. Fullbright G, Lacy ER, Büllesbach EE. The prothoracicotropic hormone bombyxin has specific receptors on insect ovarian cells. Eur J Biochem 1997; 245:774-780.
9. Geraerts WPM, Smit AB, Li KW et al. The light green cells of lymnaea: A neuroendocrine model system for stimulus-induced expression of multiple peptide genes in a single cell type. Experientia 1992; 48:464-473.
10. Schwabe C, LeRoith D, Thompson RP et al. Relaxin extracted from protozoa *(tetrahymena pyriformis)*. J Biol Chem 1983; 258:2778-2781.
11. Collier E, Watkinson A, Cleland CF et al. Partial purification and characterization of an insulin-like material from spinach and Lemna gibba G3. J Biol Chem 1987; 262:6238-6247.

CHAPTER 4

Other Mammalian and Chondrichtian Relaxins

The first relaxin sequence published was porcine relaxin and all subsequent sequences were necessarily compared with it. Rat relaxin was purified two years later,[1] and within another two years the sequence was published by John et al.[2] The sequence was significantly different from that of porcine relaxin which was surprising in light of the similarity of rat and pig insulins. Variability proved to be the rule rather than the exception for relaxin structures isolated from different animals (Fig. 4.1). Rat relaxin would eventually play a significant role in the development of structure activity relations, in particular the receptor-binding site configuration.

Meanwhile two first-year medical students decided to do a research project in the author's laboratory and their project title was 'shark relaxin'. The shark was considered, and still is by some, to be more of a living fossil than other animals, and the test question was: does the shark contain a relaxin in spite of the absence of bones, not to mention a symphysis pubis, and if so does it resemble shark insulin more than pig relaxin resembles pig insulin? After all, sharks lived 360 million years ago in the Devonian period in a form easily recognized as such, a mere 200 million years after the first appearance of fossilizable animals in the Cambrian period. Molecular biologists insist that relaxin is a duplication product of the insulin gene, and if that were true it was argued that in sharks, frozen in time as they were, the insulin and relaxin should show more homology with each other than the same pair of hormones in pigs or rats, which are "modern" animals. This idea we decided to test.

There were many jokes about male and female shark identification, but we finally decided to leave it to the Shark Fishing Club of Charleston to catch the animals and sex them properly for us. Bert

Relaxin and the Fine Structure of Proteins, by Christian Schwabe and Erika E. Büllesbach. © 1998 Springer-Verlag and R.G. Landes Company.

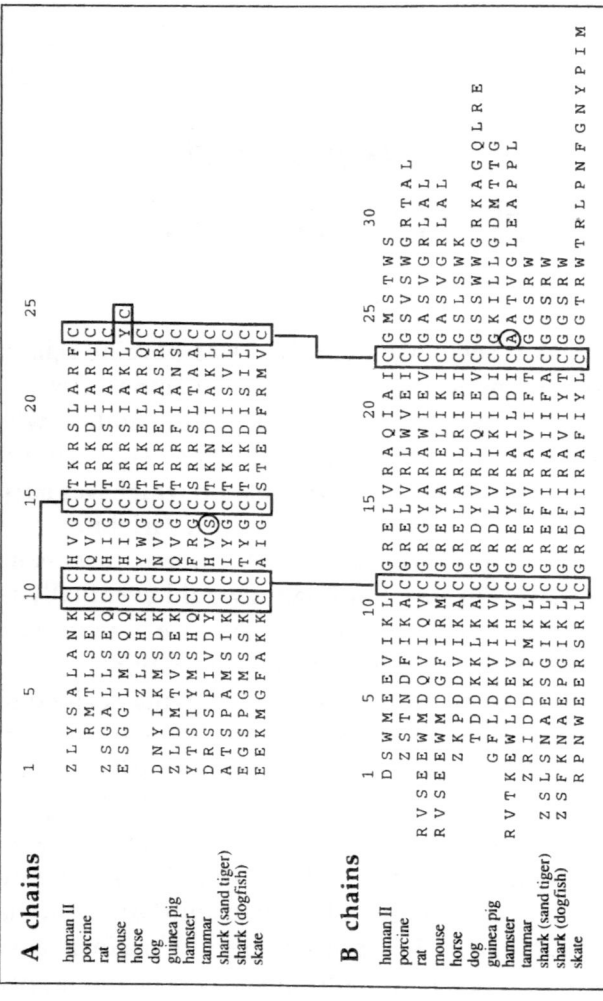

Fig. 4.1. Comparison of vertebrate relaxin sequences. The figure emphasizes the diversities of relaxin structures throughout the animal kingdom. Not included are sequences of primates (see Fig. 4.2) and sea mammals (see Fig. 4.3). Circled amino acids indicate the change of mostly constant glycine residues which are believed to be critical to protein folding. Protein sequences were reported on porcine, rat, horse, dog, sharks and skate relaxins. Sequences derived from cDNAs are human II, mouse, guinea pig, hamster, and tammar relaxins.

Daniels and Jim Reinig did a great job isolating shark relaxins, guided by nothing but a feeble crossreactivity to antiporcine relaxin antibodies.[3] In 1981 the sequence of the first chondrichtian relaxin appeared in FEBS letters.[4] While it looked neither more nor less different from pig insulin than pig relaxin did, and since shark insulin could not be found in the literature it was decided to take time off for a shark insulin sequence determination.[5] Isolating and sequencing the insulin from spiny dogfish led to the conclusion that the relaxin of cartilaginous fishes and the insulins of the same species were indeed no closer to each other than the pair of hormones was in pigs or humans. The authors' amazement concerning this observation was explained as naiveté rather than as failure of the neo-Darwinian concept of molecular evolution. This "ad hominem" suffered somewhat when it could be shown that shark relaxin differed no more from pig relaxin than pig relaxin differed from rat relaxin, for example, and that conjured up a truly awkward genealogy.

Human relaxins were deduced from their genes in 1983 and 1984.[6,7] Humans have two relaxin genes which differ from each other by 22% and each of them showed only about 40-45% homology with porcine relaxin.

Other relaxins from chondrichtians were isolated and sequence analyses showed that *Squalus acanthias* (spiny dogfish)[8] and *Raja erinacea* (skate)[9] were not significantly closer to each other than to most of the mammalian relaxins. Shortly thereafter the isolation and sequence analyses of two cetacean relaxins derived from *Balaenoptera edeni* and *Balaenoptera acutorostrata* shook up the relaxin community again. Against the background of high variability of relaxins from purportedly closely related species the relaxins of pig and whale were essentially identical (Fig. 4.2).[10] Another sea mammalian relaxin (porpoise) was isolated and sequenced by a new method involving enzymatic digestion on a sample foil followed by mass spectrometry in combination with conventional micro-Edman degradation.[11] Porpoise relaxin differed from porcine relaxin only in one amino acid residue.

Another group of animals, primates, share high sequence similarity to human relaxin. The structure of rhesus monkey relaxin, deduced from a single gene copy by Crawford et al, has 76% sequence homology to human relaxin II.[12] A similar cDNA sequence was described for baboon relaxin.[13] Other nonhuman primate genes as identified by Evans et al in the chimpanzee, orangutan, and gorilla have

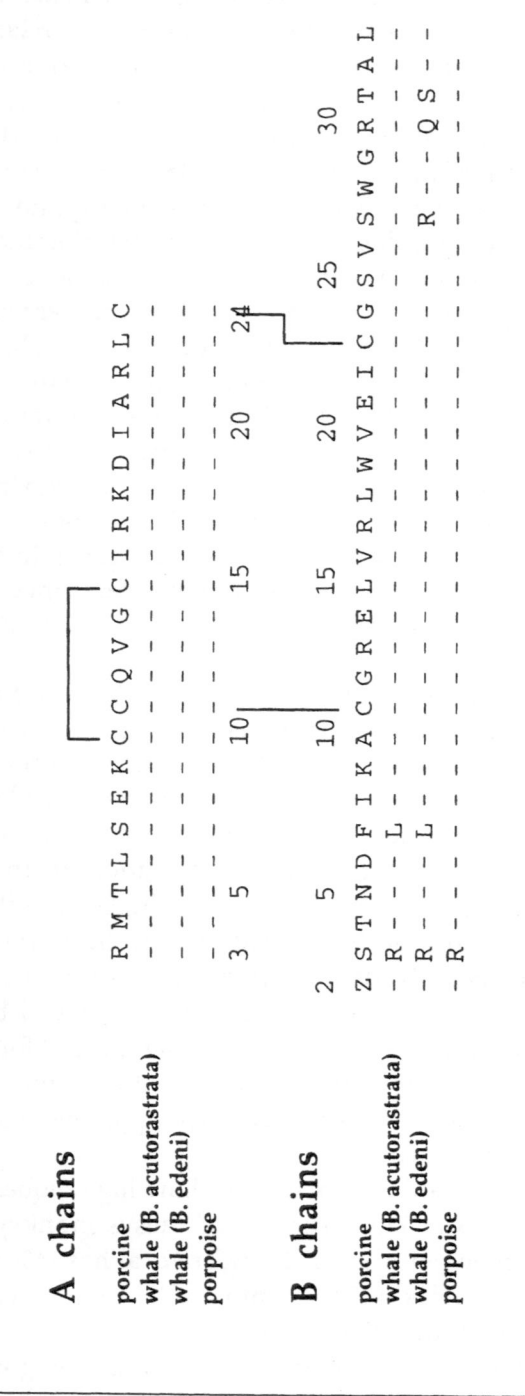

Fig. 4.2. Amino acid sequences of relaxin of sea mammals compared with porcine relaxin. To indicate the high homology between these molecules only the variable residues are shown.

two relaxin copies with sequences that show similarity to human relaxin I and human relaxin II (Fig. 4.3).[13,14]

Horse relaxin was isolated and its primary structure determined by Stewart et al,[15] followed by the isolation and sequence analysis of dog relaxin.[16] Again the structural difference between the new relaxins and the known sequences provided another example of how wide a range of primary structures can cause the same biological effect. Yet another example of high variability was provided by a relaxin cDNA isolated from guinea pig endometrium.[17] This was an important step since the guinea pig had been, for some time, the standard animal for relaxin activity measurements. Chemical synthesis and subsequent studies proved this to be a peculiar molecule and the guinea pig relaxin receptor to be equally different as regards its unusual promiscuity toward other relaxins, including that of chondrichtians.[18] Golden Syrian hamster relaxin, deduced from a cDNA by McCaslin et al, was the first relaxin in which a conserved glycine was replaced by alanine (Fig. 4.1).[19] The molecule was synthesized by us for these investigators but the bioactivity of this relaxin in a fellow rodent, the mouse, was not remarkable.

The next relaxin surprise was a mouse hormone deduced from its cDNA.[20] There was 76% homology to rat relaxin but mouse relaxin had an extra residue (tyrosine) just before the C-terminal cysteine in the A chain (Fig. 4.1). The large loop of the mouse relaxin was therefore no longer insulin-like but rather larger by one residue. The protein was synthesized according to the structure and showed a surprisingly normal helix content by CD spectroscopy.[21] The predicted consequence, i.e., looping or helix-breaking in the A chain, could not be observed and we thus learned that the disulfide bond, although limited to a 90° dihedral angle, could twist in such a way as to accommodate an extra residue in the A chain without looping.[21]

A marsupial relaxin cDNA of the tammar was recently reported to have a structurally important glycine residue replaced by a serine (Fig. 4.1).[22] The chemically synthesized tammar relaxin showed significant crossreactivity to the mouse relaxin-receptor but the affinity was less than that for porcine or human relaxin (Büllesbach and Schwabe, unpublished).

The last addition to this list is a not yet fully adopted gene product called SQ-10 which was isolated from the rabbit uterus and placenta in form of a cDNA.[23] This molecule needs to be synthesized

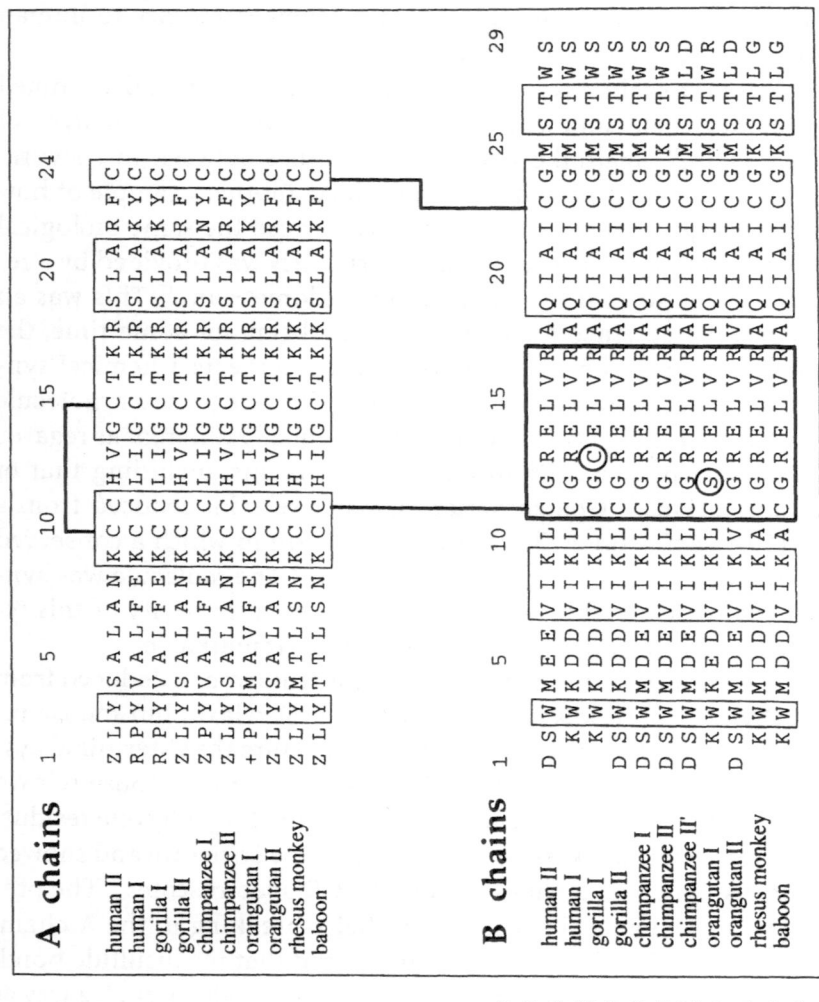

Fig. 4.3: Amino acid sequences derived from primate relaxin genes compared with the cDNA sequence of human relaxin II. In the center of the B chain the receptor-binding region is emphasized and within this conserved region two encircled residues vary. Orangutan I is probably a pseudo gene with a stop-codon in position A1 (+) and therefore the serine in position B12 is of no relevance. Gorilla I relaxin has overall seven cysteine residues with one residing in position B13, replacing an arginine residue crucial for receptor-binding.

and characterized by various bioassays and radioimmunoassays before more can be said about its nature. Not everything found in the genome is a priori functional or important for an organism and these tests can only be done when the protein is available.

The extended family of relaxin/insulin-like two-chain molecules include: the relaxin-like factor which was originally isolated in the form of its cDNA from human and porcine Leydig cells[24,25] and named accordingly the Leydig insulin-like factor (Ley I-L). Within a few months after publication of the cDNA (1994) the protein was synthesized by our new method and tested for biological activity and receptor-binding in several tissues and consequently renamed RLF for relaxin-like factor.[26] The mouse testicular RLF cDNA was obtained by Pusch et al,[27] and the synthetic protein was found to be as potent as human RLF (Büllesbach and Schwabe, unpublished).

While Adham et al believed that the RLF message is Leydig cell-specific,[24] Tashima et al, on the basis of hybridization experiments, suggested that the human corpus luteum and trophoblast also express this molecule.[28] That RLF is expressed in females was clearly demonstrated in ruminants, cow and sheep, where high expression was observed in the ovary during cycling and pregnancy.[29,30]

Another member of this two-chain insulin/relaxin-like family of molecules was discovered in form of a cDNA by Chassin et al (1995) in human placenta and the gene locus was named INSL-4.[31] This molecule has distinctly less relaxin-like qualities and falls short of insulin likeness because of an interrupted hydrophobic core region. After the synthesis of the protein (Büllesbach and Schwabe, unpublished) this assumption was verified by CD measurements which clearly showed the lack of helix content of INSL-4 in water.

A family of proteinaceous factors of about the same overall structure called bombyxin was isolated from the lepidopteran *Bombyx mori*. The first sequence information on bombyxin was published in the mid 1980s by Nagasawa et al in *Science* and in *Proceedings of the National Academy of Sciences*.[32,33] The group also synthesized bombyxin II and bombyxin IV in 1992 by chemical methods similar[34,35] to those developed for our synthesis of human relaxin in 1991.[36] Meanwhile insulin-like sequences from other insects were published, including the ailanthus moth, *Samia cynthia*,[37] the hornworm, *Agrius convoluvuli*,[38] and the locust, *Locusta migratoria*.[39]

The parade of relaxin/insulin-like factors continues with a snail-derived insulin-related peptide, a molecule that has four instead of the usual three disulfide bonds[40] and thus presents a significantly different problem during chemical synthesis. The function of this hormone is not known but there are suggestions that it might act insulin-like in the snail.[41]

A very recent discovery is a relaxin-like molecule from the alkaline gland of the stingray *Daysiatis sabina*. The molecule has several of the important features of relaxin and is the first of this genre with an oligosaccharide chain linked to an asparagine residue in the C-terminal region of the B chain.[42] The alkaline gland in the stingray is, to some extent, an equivalent of the mammalian prostate which prompted the search for a relaxin-like molecule in the alkaline gland fluid. To call this molecule "raylaxin" seemed irresistable to all but the editor of *Biochemistry* who insisted on stingray relaxin. The molecule did not show the expected sperm mobility enhancement activity ascribed to human prostate relaxin so that no comment can be made as to its functions. Furthermore, while the human male and female relaxin are products of the same gene, *D. sabina* relaxin has little homology to the female relaxin of the skate *Raja erinacea*.[42]

High-technology literature-based sequence searches and rapid cDNA screening techniques are likely to reveal many more of these two-chain factors which could act as autocrines or paracrines in the tissues of origin. One might suspect that beyond the known hormones there are many more factors required to support all of the biological functions or the fine-tuning of metabolism of which we are aware.

References

1. Sherwood OD. Purification and characterization of rat relaxin. Endocrinology 1979; 104:886-892.
2. John M, Borjesson BW, Walsh JR et al. Limited sequence homology between porcine and rat relaxins: Implication for physiological studies. Endocrinology 1981; 108:726-729.
3. Reinig JW, Daniel LN, Schwabe C et al. Isolation and characterization of relaxin from the sand tiger shark *(Odontaspis taurus)*. Endocrinology 1981; 109:537-543.
4. Gowan LK, Reinig JW, Schwabe C et al. On the primary and tertiary structure of relaxin from the sand tiger shark *(Odontaspis taurus)*. FEBS Lett 1981; 129:80-82.
5. Bajaj M, Blundell TL, Pitts JE et al. Dogfish insulin. Primary structure, conformation and biological properties of an elasmobranchial insulin. Eur J Biochem 1983; 135:535-542.
6. Hudson P, Haley J, John M et al. Structure of a genomic clone encoding biologically active human relaxin. Nature 1983; 301:628-631.

7. Hudson P, John M, Crawford R et al. Relaxin gene expression in human ovaries and the predicted structure of a human preprorelaxin by analysis of cDNA clones. EMBO J 1984; 3:2333-2339.
8. Büllesbach EE, Gowan LK, Schwabe C et al. Isolation, purification, and sequence of relaxin from spiny dogfish (*Squalus acanthias*). Eur J Biochem 1986; 161:335-341.
9. Büllesbach EE, Schwabe C, Callard IP. Relaxin from an oviparous species, the skate (*Raja erinacea*). Biochem Biophys Res Commun 1987; 143:273-280.
10. Schwabe C, Büllesbach EE, Heyn H et al. Cetacean Relaxin: Isolation and Sequence of Relaxins from *Balaenoptera acutorostrata* and *Balaenoptera edeni*. J Biol Chem 1989; 264:940-943.
11. Woods AS, Cotter RJ, Yoshioka M et al. Enzymatic digestion on the sample foil as a method for sequence determination by plasma desorption mass spectrometry: the primary structure of porpoise relaxin. Int J Mass Spectrometry Ion Processes 1991; 111:77-88.
12. Crawford RJ, Hammond VE, Roche PJ et al. Structure of rhesus monkey relaxin predicted by analysis of the single-copy rhesus monkey relaxin gene. J Mol Endocrinol 1989; 3:169-174.
13. Evans BA, Fu P, Tregear GW. Characterization of primate relaxin genes. Endocrine J 1994; 2:81-86.
14. Evans BA, Fu P, Tregear GW. Characterization of two relaxin genes in the chimpanzee. J Endocrinol 1994; 140:385-392.
15. Stewart DR, Nevis B, Hadas E et al. Affinity purification and sequence determination of equine relaxin. Endocrinology 1991; 129:375-383.
16. Stewart DR, Henzel WJ, Vandlen R. Purification and sequence determination of canine relaxin. J Prot Chem 1992; 11:247-253.
17. Lee YA, Bryant-Greenwood GD, Mandel M et al. The complementary desoxyribonucleic acid sequence of guinea pig endometrial prorelaxin. Endocrinology 1992; 130:1165-1172.
18. Büllesbach EE, Steinetz BG, Schwabe C. Synthesis and biological properties of guinea pig relaxin. Endocrine 1994; 2:1115-1120.
19. McCaslin RB, Renegar RH. Determination of the prorelaxin nucleotide sequence and expression of prorelaxin messenger ribonucleic acid in the golden hamster. Biol Reprod 1995; 53:454-461.
20. Evans BA, John M, Fowler KJ et al. The mouse relaxin gene: nucleotide sequence and expression. J Mol Endocrinol 1993; 10:15-23.
21. Büllesbach EE, Schwabe C. Mouse relaxin: Synthesis and biological activity of the first relaxin with an unusual crosslinking pattern. Biochem Biophys Res Commun 1993; 196:311-319.
22. Parry LJ, Rust W, Ivell R. Marsupial relaxin—complementary deoxyribonucleic acid sequence and gene expression in the female and male tammar wallaby, *Macropus eugenii*. Biol Reprod 1997; 57:119-127.
23. Fields P, Kondo S, Tashima L et al. Expression of SQ10 (a preprorelaxin-like gene) in the pregnant rabbit placenta and uterus. Biol Reprod 1995; 53:1139-1145.

24. Adham IM, Burkhardt E, Benahmed M et al. Cloning of a cDNA for a novel insulin-like peptide of the testicular Leydig cells. J Biol Chem 1993; 268:26668-26672.

25. Burkhardt E, Adham IM, Brosig B et al. Structural organization of the porcine and human genes coding for a Leydig cell-specific insulin-like peptide (LEY I-L) and chromosomal localization of the human gene (INSL3). Genomics 1994; 20:13-19.

26. Büllesbach EE, Schwabe C. A novel Leydig cell cDNA-derived protein is a relaxin-like factor (RLF). J Biol Chem 1995; 370:16011-16015.

27. Pusch W, Balvers M, Ivell R. Molecular cloning and expression of the relaxin-like factor from the mouse testis. Endocrinology 1996; 137:3009-3013.

28. Tashima LS, Hieber AD, Greenwood FC et al. The human Leydig insulin-like (hLEY I-L) gene is expressed in the corpus luteum and trophoblast. J Clin Endocrinol Metab 1995; 80:707-710.

29. Bathgate R, Balvers M, Hunt N et al. Relaxin-like factor gene is highly expressed in the bovine ovary of the cycle and pregnancy— Sequence and messenger ribonucleic acid analysis. Biol Reprod 1996; 55:1452-1457.

30. Roche PJ, Butkus A, Wintour EM et al. Structure and expression of Leydig insulin-like peptide mRNA in the sheep. Mol Cell Endocrinology 1996; 121:171-177.

31. Chassin D, Laurent A, Janneau JL et al. Cloning of a new member of the insulin gene superfamily (INSL4) expressed in human placenta. Genomics 1995; 29:465-470.

32. Nagasawa H, Kataoka H, Isogai A et al. Amino-terminal amino acid sequence of the silkworm prothoracicotropic hormone: Homology with insulin. Science 1984; 226:1344-1345.

33. Nagasawa H, Kataoka H, Isogai A et al. Amino acid sequence of a prothoracicotropic hormone of the silkworm *Bombyx mori.* Proc Natl Acad Sci USA 1986; 83:5840-5843.

34. Maruyama K, Nagata K, Tanaka M et al. Synthesis of bombyxin-IV, an insulin superfamily peptide from the silkworm, *Bombyx mori,* by stepwise and selective formation of three disulfide bridges. J Prot Chem 1992; 11:1-12.

35. Nagata K, Maruyama K, Nagasawa H et al. Bombyxin-II and its disulfide bond isomers: Synthesis and activity. Peptides 1992; 13:653-662.

36. Büllesbach EE, Schwabe C. Total synthesis of human relaxin and human relaxin derivatives by solid phase peptide synthesis and site-directed chain combination. J Biol Chem 1991; 266:10754-10761.

37. Kimura-Kawakami M, Iwami M, Kawakami A et al. Structure and expression of bombyxin related peptide genes of the moth *Samia cynthia ricini.* Gen Comp Endocrinol 1992; 86:257-268.

38. Iwami M, Furuya I, Kataoka H. Bombyxin-related peptides, cDNA structure and expression in the brain of the hornworm *Agrius convoluvuli.* Insect Biochem Mol Biol 1996; 26:25-32.

39. Hetru C, Li KW, Bulet P et al. Isolation and structural characterization of an insulin-related molecule, a predominant neuropeptide from *Locusta migratoria*. Eur J Biochem 1991; 201:495-499.
40. Geraerts WPM, Smit AB, Li KW et al. The light green cells of lymnaea: a neuroendocrine model system for stimulus-induced expression of multiple peptide genes in a single cell type. Experientia 1992; 48:464-473.
41. Geraerts WP. Neurohormonal control of growth and carbohydrate metabolism by the light green cells in Lymnaea stagnalis. Gen Comp Endocrinol 1992; 86:433-444.
42. Büllesbach EE, Schwabe C, Lacy ER. Identification of a glycosylated relaxin-like molecule from the male Atlantic stingray, *Dasyatis sabina*. Biochemistry 1997; 36:10735-10741.

CHAPTER 5

Isolation and Sequence Analysis of Porcine Relaxin

The fetuses of a sow would have to be as long as the width of a man's fist before the ovaries would contain significant amounts of relaxin. About 100 kg of ovaries collected under contract in an abattoir had to be discarded because the collectors had ignored the order to do this rough measurement on the fetuses. A little inconvenience had turned them into endocrinologists which made them decide that the large yellow cysts that often occurred in pig ovaries were a better sign of pregnancy. This costly mistake leaves something to be said for specialization.

Porcine relaxin is most efficiently extracted by the patented method of Doczi with 70% acetone and 0.15 normal HCl.[1] The first pellet obtained by low-speed centrifugation at this point is discarded and the supernatant brought up to 90% in cold acetone and kept for 48 h at 4°C. By that time the relaxin, among other things, has precipitated out and forms a relatively firm film from which the acetone can be siphoned off carefully. The pellet is then extracted with distilled water and dialyzed against two changes of water followed by 50 mM ammonium acetate buffer at pH 5.5. After 48 hours of dialysis the suspension is cleared by slow speed centrifugation and the supernatant applied to ion exchange chromatography on a CM cellulose column equilibrated with 50 mM ammonium acetate at pH 5.5. This is essentially the method designed by Sherwood which yields several relaxin-containing fractions, called CM-a, CM-a', and CM-B according to their position of elution (Fig. 5.1).[2] The procedure was a significant step forward because relaxin could now be defined as a family of compounds with 3000 units of activity/mg and one unit as the amount that would cause half of a group of guinea pigs to show

Relaxin and the Fine Structure of Proteins, by Christian Schwabe and Erika E. Büllesbach. © 1998 Springer-Verlag and R.G. Landes Company.

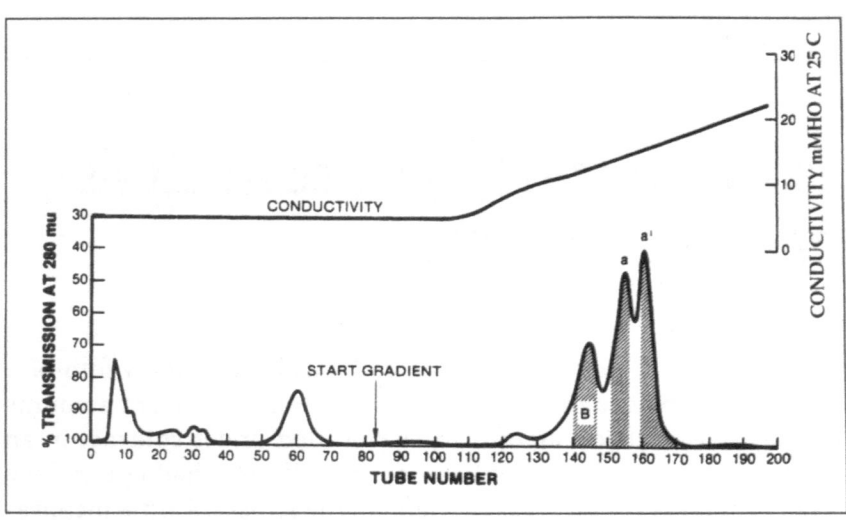

Fig. 5.1. Ion exchange chromatography on CM-cellulose in 80 mM ammonium acetate/ acetic acid pH 5.5. Porcine relaxin was eluted with a linear NaCl gradient. Three porcine relaxin fractions (shaded) were of equivalent potency. Reprinted with permission from: Sherwood, OD, O'Byrne EM. Arch Biochem Biophys 1974; 160:185-196.

relaxation in the symphysis pubis. The guinea pig assay, cumbersome and difficult as it is, has been largely replaced by the mouse pubic symphysis relaxation assay where one microgram is sufficient to cause complete ligament formation. Still the CM cellulose fractions were not sufficient for the type of work we planned to do with relaxin. Only if one starts out with a single component can a targeted modification lead to a defined derivative that can be separated from unreacted starting material and byproducts. In order to obtain a single compound to be called porcine relaxin we first had to know the structural basis of the observed heterogeneity.

The first purification was done under much more humble conditions, starting with a relaxin preparation that was given to investigators upon request in milligram amounts by the National Institutes of Health (NIH) hormone distribution center. That was a fine arrangement but the shipments obtained varied from a white powder to a brownish caramel, indicating that relaxin in this preparation must have been a minor component. Contact with the distribution officer did not increase our supply measurably but provided as an aside the reassuring information that somebody else in the field was about to finish the relaxin sequence work! Considering the fact that we did not even have automatic sequence equipment, had little experience with the sequencing procedures and no good source for starting material, we felt like Garrison Keillor when only rhubarb pie can save one from despair and humiliation. These are the moments when one asks oneself, why relaxin?

The answer is brief, simple, and as compelling as most stories in science; it began with a Gordon Research Conference where a dramatic effect of relaxin on connective tissues was reported together with various comments about the esthetically displeasing relaxin preparation that was issued by NIH. Over cocktails we promised to purify the material for the speaker, retain some for sequence analysis, and send homogeneous material back to him for his physiological studies. A few weeks later the curiosity about relaxin per se overtook the interest in connective tissue metabolism and that was it.

The next part of the story sounds more like a fairy tale. Within a day of these events the late vice president of our university called and asked whether we still wanted the instrument we had been talking about so long. He had found funds with a two-weeks spending limit and that, if we could get the instrument into the laboratory and invoiced by that time, he would pay for it. The Beckman team could be persuaded to modify an 890C that was ready to be shipped to Moscow and to put it into our laboratory within the prescribed (record) time. That was a great step forward but while we were still far behind, to see this beautiful piece of precision engineering and hear it clicking away to the rhythmic hum of two gigantic vacuum pumps gave us a tremendous boost.

In fact at that time the first direct purification of relaxin from pregnant hog ovaries had been published by Sherwood and O'Byrne[2] who worked in Dr. Steinetz's lab and Dr. Steinetz, a senior investigator at Ciba Geigy, provided us with the first purified relaxin to get our sequence work started. It was also clear at that time that relaxin was a two-chain molecule held together by disulfide crosslinks and that for proper sequencing the molecule needed to be purified, reduced, and the A and B chains separated cleanly. Sherwood and his co-workers had separated the chains after reduction and S-carboxymethylation by size separation on a Sephadex G50 column in guanidine-HCl at pH 5.0. The A chain showed a large absorption peak at 230 nm and no absorbance at 280 nm and eluted after the larger B chain. The B chain gave a much lower signal at 230 nm but absorbed UV light at 280 nm (Fig. 5.2).[2] The relative yield of the

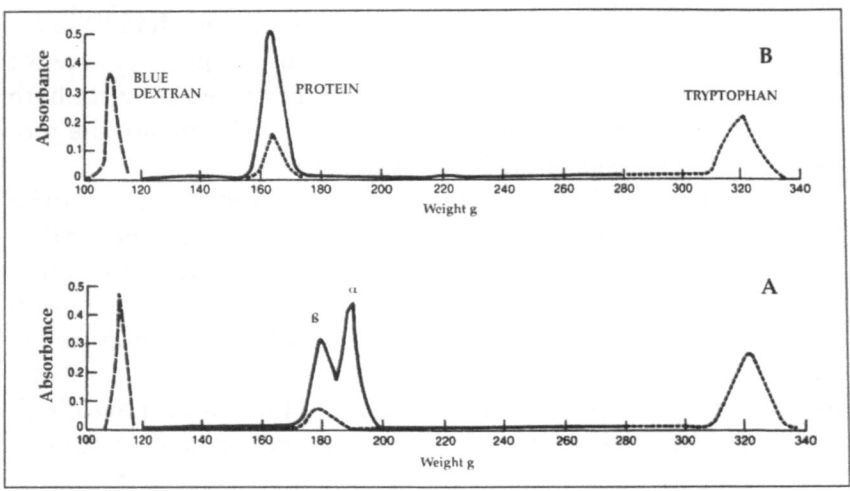

Fig. 5.2. Gelfiltration on Sephadex G50 superfine in 10 mM ammonium acetate/acetic acid pH 5.0 in 6 M guanidinium chloride. Upper panel: intact porcine relaxin, lower panel porcine relaxin after reduction and carboxymethylation. The column was standardized with blue dextran and tryptophan to indicate the exclusion volumes of the column. Protein was determined by UV absorbance at 280 nm (dashed lines) and 230 nm (solid line). Reprinted with permission from: Sherwood, OD, O'Byrne EM. Arch Biochem Biophys 1974; 160:185-196.

chains implied that either the B chain was irreversibly adsorbed to the column or the relaxin had a ratio of two A chains for each B chain. The separation required a 200 cm column to run for several days and was therefore neither suited for our temperament nor for the time constraints of competitive work.

Relaxin remained difficult to handle and once we had reduced it in a guanidinium chloride solution at pH 8.6 and then diluted with acetic acid it always precipitated out. Insolubility is a bad problem to deal with and was perceived as such until we did an experiment involving alkylation of the SH groups with tritiated iodoacetamide. The same problem occurred but this time the precipitate appearing upon the acidification was centrifuged and the supernatant tested for radioactivity. The clear supernatant contained pure relaxin A chain; serendipity had saved us months of headaches. Upon sequencing, the A chain proved to be insulin-like as far as the distribution of cysteine residues was concerned but significantly different otherwise. We must be catching up!

No sequence of relaxin had appeared in the literature in spite of the hormone officers ominous remarks, and since an endocrine society meeting was about to be held in San Francisco and no relaxin abstract had been listed, we decided to present our results which were interesting because of the insulin likeness of the A chain even though the B chain (for cause) had still resisted all attempts at sequencing.

Something extraordinary happened in San Francisco. As soon as the paper was opened for discussion a young lady jumped to her feet declaring in a jubilant voice that we were to be congratulated for having confirmed her data. She then proceeded to show a few slides one of which, bearing the brown stains of great age, was depicting a partial sequence of the A chain of relaxin; the sequence had already been finished of course. There were a few moments of silence in the room, followed by murmuring in the audience, and finally the question from the speaker concerning the whereabouts of the relevant publication. Her statement that it was not yet published was greeted with another round of background murmur.

In one moment the world collapsed and in the next one everything is in place again. These are some of the embellishments of a science career which make life interesting and enjoyable if one is winning. At the day of the meeting our paper had appeared in press.

Back home it became quite clear that the B chain had a blocked N-terminal end which could be due to either a pyrrolidone carboxylic acid (Pca) or an N-acetyl group. The arrival of this realization coincided with the arrival of Dr. Ken McDonald as a new faculty member and professor of the Department of Biochemistry. One of the authors had pushed coincidence along a little bit by recruiting him from sunny California to sunny Charleston with a 50/50 mixture of half-truths and half-lies called the recruiting cocktail. Finally, academic arguments prevailed—we can raise poinsettias in Sunnyvale—well we do not because they grow wild in Charleston.

In his career Dr. McDonald had debunked the claims of many years of cathepsin C work which had led to erroneous conclusions that this enzyme acts as an endopeptidase, a protease that splits internal bonds in a polypeptide chain. He found that the enzyme has a novel catalytic mode of action, causing the removal of two amino acids at a time from the N-terminal end and thus degrading the chain into dipeptides as long as every clip exposes another removable dipeptide.[3] Consequently the enzyme was named dipeptidylamino peptidase-I (DAP-I). Would it remove the blocked amino terminus from the B chain? Within a few days experiments on intact relaxin were performed which turned out negative. Chronologically these experiments were performed a few weeks before the chemical sequence data and consequently there was no explanation for the lack of activity of this highly touted enzyme. The discovery of an arginine at the N-terminal end of the A chain provided the explanation. DAP-I is unable to digest a protein chain that begins with either arginine or lysine. The B chain however was refractory because there was no free amino group which is required for both Edman degradation and dipeptide amino peptidase activity. Eventually we modified the intact relaxin with an N-carboxyanhydride derivative of tyrosine which can be prepared by carefully diffusing phosgene through a solution of tyrosine in the presence of sulfuric acid. The crystalline product can be stored at 80 degrees in a desiccator. Added to a relaxin solution in dimethylformamide this activated amino acid would add to the primary amino group and in this case modify the relaxin A chain such as to become a substrate for DAP-I, i.e., to begin with tyrosine, i.e., Tyr-Arg-Met-Thr-Leu-Ser- instead of Arg-Met-Thr-Leu-Ser- and consequently the new enzyme produced the fragments Tyr-Arg, Met-Thr, and Leu-Ser. These results confirmed our sequence analysis, rescued the reputation of DAP-I and thus made everybody happy.

The B chain remained stubborn. The thin-layer chromatography (tlc) technology eventually allowed us to detect pyroglutamic acid (Pca) on silica gel plates. The enzyme called appropriately Pca-peptidase was incubated with the intact relaxin and the reaction mixture was spotted as a function of time on a tlc plate to show the appearance of pyroglutamic acid but the removal of the blocking group was very inefficient.[4]

This interlude was thought to have prepared the path for a B chain sequence. The Pca-peptidase would not remove the pyroglutamic acid residue from the intact B-chain. Unlike trypsin, the Pca-peptidase requires complete solubility of the substrate. There was no choice but to digest the B-chain with trypsin, to separate the peptides and to sequence the fragments.[5] The B chain pellet obtained from reduction and S-alkylation of intact relaxin was washed a few times in diluted acetic acid, spun down and taken up again in 50 mM ammonium bicarbonate which solubilized the chain sufficiently for protease digestion. The peptides were separated by tlc or HPLC and analyzed. Unfortunately, the Beckman 890C sequencer was not at its best with small peptides even in the presence of polybrene, a polymer-containing quarternary ammonium groups used to retain proteins or peptides in the sequencer cup. A relatively small peptide will wash out severely and no matter how much chain we started with, toward the end the signals became rather feeble. Only matching results of amino acid analysis and sequence analysis made it possible to define a complete sequence. Ambiguous results called for carefully designed carboxypeptidase A experiments to determine the C-terminal residues which were later found to be due to heterogeneity.[6,7] In addition tryptic fragments of the succinylated B chain were produced and separated. The combined information was sufficient to organize the peptides in the proper sequence.[5]

There was no doubt that the insulin similarity continued and a tryptic digest of unreduced relaxin provided evidence for the insulin-like crosslinks.[8]

Highly purified relaxin (5 mg) was dissolved in N-ethylmorpholine buffer at pH 8.5 and incubated for 4 hours with 50 μg of trypsin at 37°C. The cysteine-containing peptides were identified after separation by high performance liquid chromatography (HPLC) on a reverse-phase C_{18} column (Fig. 5.3). Peptide 2 upon Edman degradation yielded Ala during the first step, nothing during the second, and Glu and Gly followed by Val and Arg during the second and third cycle respectively. Accordingly Cys A9 was linked to Cys B10 and A8 to A13. This established the crosslinks for porcine relaxin.

This was the first time that the crosslinks were established by experiment; much later, combining liquid sequencing, mass spectrometry and chemical synthesis of tryptic fragments, the crosslinks were confirmed as originally recorded.[9] For all other relaxins crosslinks were assumed to be the same which is correct so far with the exception of mouse relaxin. The point was eventually proven by synthetic work so that one can state today, with some confidence, that all known relaxins except that of the mouse have the same insulin-like disulfide crosslinking pattern. The B chain sequence was submitted for publication but before it could appear in print our public relations office had been bleeding this information to the wire service, and like mushrooms articles on relaxin in pregnant sows popped up in newspapers in many places in the country. Our competitors learned about the total relaxin sequence a few days before the report appeared in the professional journal. Relaxin had been known as a factor or a biological activity for so long that many

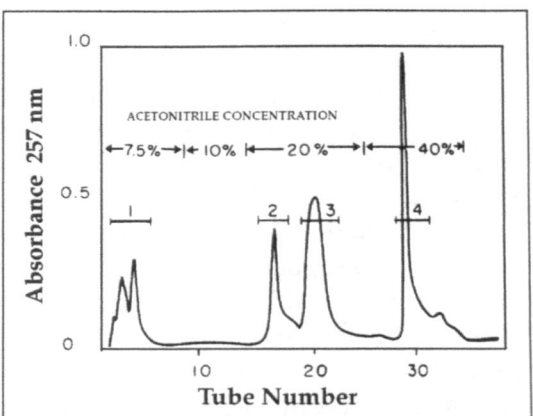

Fig. 5.3. High performance liquid chromatography of tryptic peptides derived from native porcine relaxin. The acetonitrile gradient increased stepwise and peaks were collected as indicated. Peptides 2 and three contained cystine and were further investigated. Reprinted with permission from: Schwabe C, McDonald, JK. Science 1977; 197:914-915.

speculations were circulating concerning the nature of this molecule and therefore the structural work was greeted with an unusual degree of interest. Today the sequence of a small protein would barely raise eyebrows. Discoveries seem to run on an independent schedule; running faster engenders painful rejection, too slow yields nothing.

This little bit of drama also carries the message not to be dissuaded by unfavorable odds. Competing with larger groups is not hopeless but it takes the kind of sacrifice that is not generally popular.

The greatest reward, however, came when one of the deans of relaxinologists, Dr. Roy Greep, wrote a letter to the authors conveying congratulatory thoughts ending with the statement "and there shall be a toast and fireworks to let the spirit of Hisaw know: t'is done"; (Hisaw had passed away a year before this event).

The large scale of isolation of B29 porcine relaxin, which can be repeated according to the schematic given below (Fig. 5.4), was also used to isolate the side-fractions which upon analysis made it clear that the heterogeneity of porcine relaxin was due to biological processing.[7] In contrast to insulin processing, the proteolytic activation of relaxins from the prohormones is very inaccurate. The A chains always had the same C termini but varied in length at the N-terminal ends whereas the B chain had most of its variation at the C-terminal end. Since the C peptide or the connecting peptide runs from the C-terminal end of the B chain to the N-terminal end of the A chain, inaccurate processing would produce heterogeneity in these two places. The major split near the N-terminal end of the A chain was dictated by four basic amino acids[10] but subsequent processing

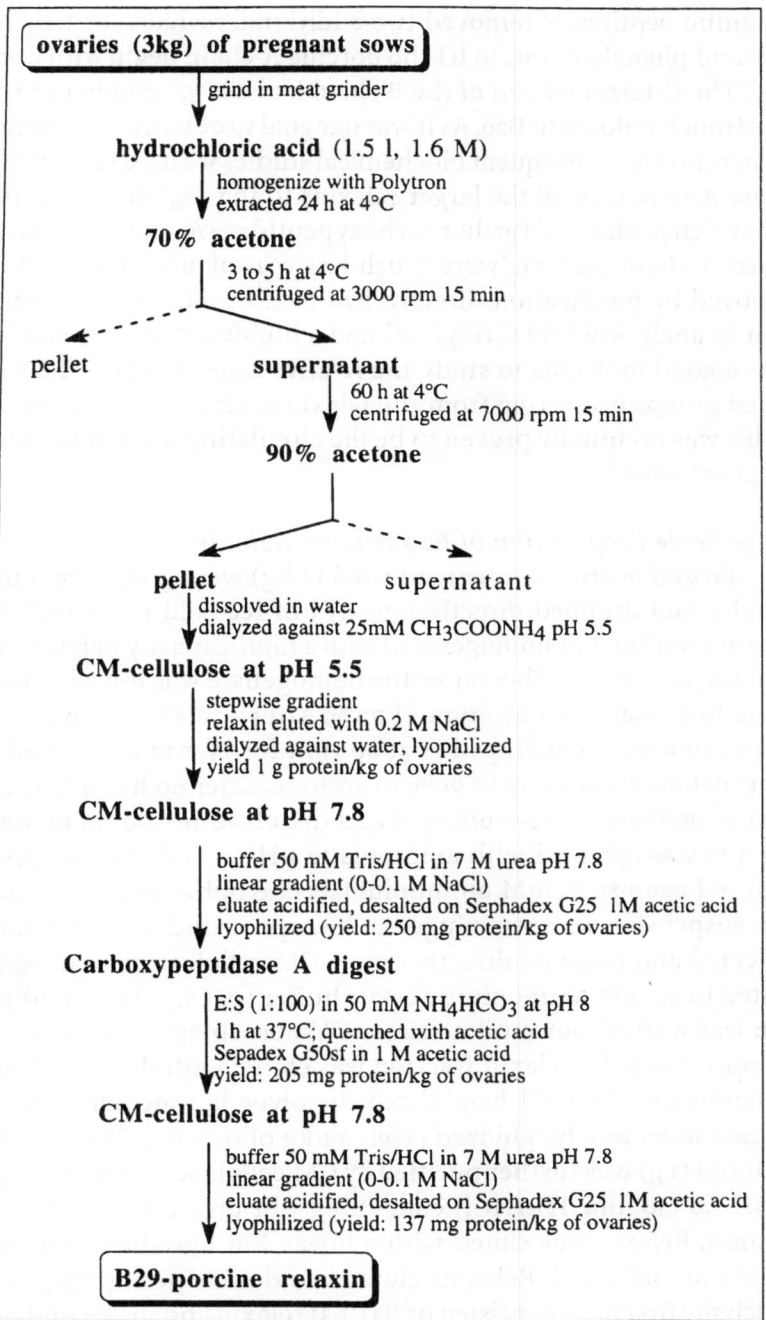

ovaries (3kg) of pregnant sows

grind in meat grinder

hydrochloric acid (1.5 l, 1.6 M)

homogenize with Polytron
extracted 24 h at 4°C

70% acetone

3 to 5 h at 4°C
centrifuged at 3000 rpm 15 min

pellet   **supernatant**

60 h at 4°C
centrifuged at 7000 rpm 15 min

90% acetone

pellet   supernatant

dissolved in water
dialyzed against 25mM CH₃COONH₄ pH 5.5

CM-cellulose at pH 5.5

stepwise gradient
relaxin eluted with 0.2 M NaCl
dialyzed against water, lyophilized
yield 1 g protein/kg of ovaries

CM-cellulose at pH 7.8

buffer 50 mM Tris/HCl in 7 M urea pH 7.8
linear gradient (0-0.1 M NaCl)
eluate acidified, desalted on Sephadex G25 1M acetic acid
lyophilized (yield: 250 mg protein/kg of ovaries)

Carboxypeptidase A digest

E:S (1:100) in 50 mM NH₄HCO₃ at pH 8
1 h at 37°C; quenched with acetic acid
Sepadex G50sf in 1 M acetic acid
yield: 205 mg protein/kg of ovaries

CM-cellulose at pH 7.8

buffer 50 mM Tris/HCl in 7 M urea pH 7.8
linear gradient (0-0.1 M NaCl)
eluate acidified, desalted on Sephadex G25 1M acetic acid
lyophilized (yield: 137 mg protein/kg of ovaries)

B29-porcine relaxin

Fig. 5.4. Purification of porcine relaxin starting with ovaries of pregnant pigs and the generation of B29-relaxin a monocomponent porcine relaxin.

by amino peptidases removed two additional residues, namely leucine and phenylalanine, to let the porcine A chain begin with arginine.[7] The C-terminal end of the B chain had a large number of variants from B32 down to B26. As it was our goal to obtain a homogenous preparation for subsequent biochemical studies we used carboxypeptidase A to reduce all the larger relaxins to the arginine in position B29 which prohibited further carboxypeptidase A action. The N-terminal A chain variants were much less conspicuous and could be removed by purification directly. B29 relaxin was a single component in analytical HPLC (Fig. 5.6) and a highly active hormone and thus a good molecule to study the relative importance of the functional groups projecting from the relaxin surface at various points. It also was eventually proven to be the circulating form of relaxin in pregnant sows.[11]

Large Scale Preparation of B29 Porcine Relaxin[7]

Frozen ovaries of pregnant sows (3 kg) were ground in a meat grinder and dropped directly into 1.5 l of ice-cold 1.6 M HCl. The mixture was further homogenized with a high-capacity Polytron and kept for 24 h at 4°C. Thereafter the homogenate was mixed with acetone to a final concentration of 70%. After 3 to 5 hours at 4°C the suspension was centrifuged at 3000 rpm for 15 min at 4°C and the supernatant brought up to 90% in acetone. After 60 h at 4°C the resulting precipitate was collected and dissolved in 500 ml of water. The pH was adjusted with ammonia to pH 5.5 and the suspension dialyzed against 25 mM ammoniumacetate buffer pH 5.5 for 24 h. The suspension was centrifuged at low speed and the supernatant collected and pumped directly onto a CM cellulose column equilibrated in 50 mM ammonium acetate buffer (pH 5.5). Unbound protein was washed out until no more UV-absorbing material left the column. Then the relaxin fraction was eluted with the same buffer containing 0.2 M NaCl, lyophilized, dissolved in water and dialyzed against water and lyophilized (yield 1g/kg of ovaries). The resulting material (1 g) was further purified by CM cellulose chromatography at pH 7.8 (50 mM Tris/HCl) containing 7 M urea using a 3 x 25 cm column. Relaxin was eluted with a linear NaCl gradient from 0 to 0.1 M (500 ml each). Relaxins eluted in a double peak (Fig. 5.5a) of which the first peak consisted of B32/B31 relaxin and the second peak of B29 relaxin. The double peak was pooled and relaxin was recovered within 24 h of the chromatography by desalting on Sephadex

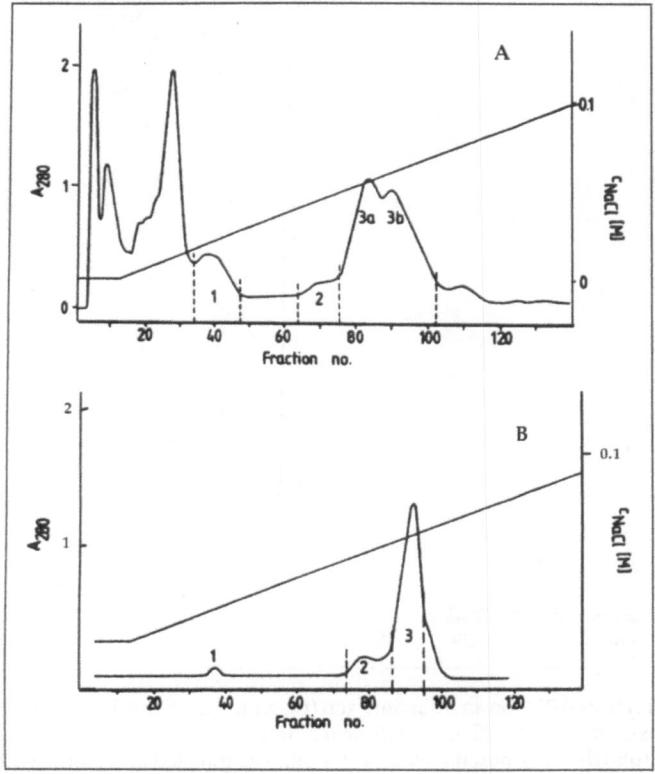

Fig. 5.5. Ion exchange chromatography on CM cellulose at pH 7.8 (column 3 cm x 25 cm; buffer 7 M urea, 0.05 M Tris/HCl pH 7.8; gradient 0-0.1M NaCl, 500 ml each, flow rate 50 ml/h, fractions 6 ml). A) porcine relaxin. The double peak (3a and 3b) was pooled, desalted, digested with carboxypeptidase A, and gelfiltered on Sephadex G-50 sf. B) rechromatography of the resulting product. Reprinted with permission from: Büllesbach EE, Schwabe C. Biochemistry 1985; 24:7717-7722.

G25 in 1 M acetic acid, followed by lyophilization. This mixture of relaxins (250 mg) was dissolved in 4 ml of 50 mM ammonium hydrogen carbonate and incubated with a freshly prepared suspension of carboxypeptidase A (2.5 mg in 1 ml of 1 M NaCl) for 1 h at 37°C. The sometimes cloudy solution was acidified with 500 µl glacial acetic acid and the resulting clear solution separated by size on Sephadex G50-sf in 1 M acetic acid in the presence of 0.15 M NaCl. The relaxin-containing fractions were pooled and desalted on Sephadex G25 in 1 M acetic acid and lyophilized. The product was rechromatographed on a freshly packed CM-cellulose column at pH 7.8 under the same

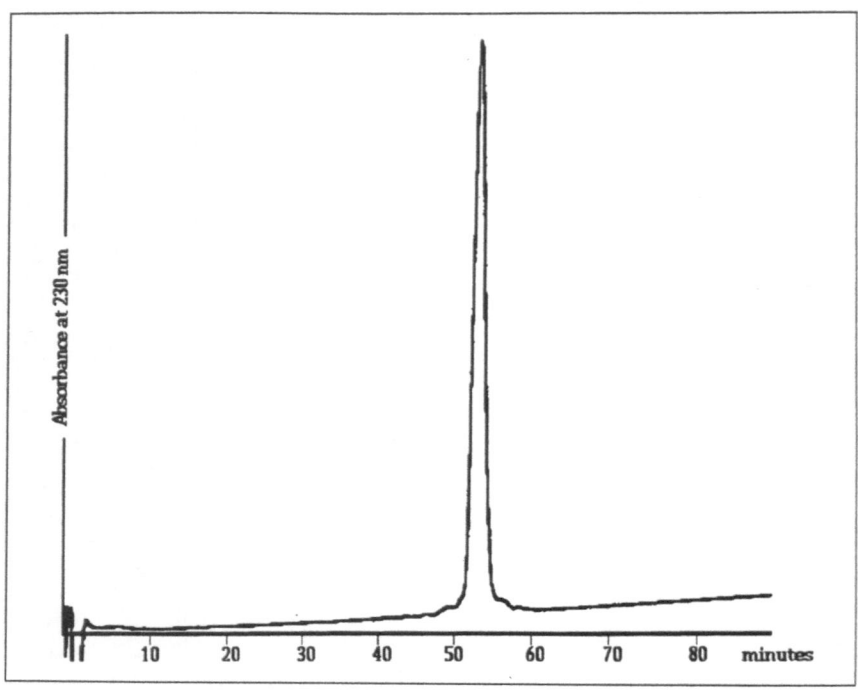

Fig. 5.6. Analytical HPLC on an Aquapore 300 (C$_8$, 2.1 mm x 30 mm) column. The solvent system consisted of 0.1% trifluoroacetic acid in water (A) and 0.1% trifluoroacetic acid in 80% acetonitrile. B29-porcine relaxin (1-2 µg) was injected and eluted with a linear gradient from 25% B to 45% B over 90 minutes at a flow rate of 100 µl/min. The column was washed with 100% B for 5 minutes before equilibrating with 25% in preparation for the next run.

conditions as described above. The chromatogram (Fig. 5.5b) showed a main peak with a small shoulder in front and the back. The less positively charged impurity was identified as A0-phenylalanyl-relaxin. B29-porcine relaxin was obtained in a yield of 137 mg/kg of ovaries.

References

1. Doczi J. Process for the extraction and purification of relaxin. US patent 3 096 246: 1963.
2. Sherwood OD, O'Byrne EM. Purification and characterization of porcine relaxin. Arch Biochem Biophys 1974; 160:185-196.
3. McDonald JK, Zeitman BB, Reilly TL et al. New observations on the substrate specificity of cathepsin C (dipeptidyl aminopeptidase I). J Biol Chem 1969; 244:2693-2709.

4. Schwabe C, McDonald JK. Demonstration of a pyroglutamyl residue at the N terminus of the B chain of porcine relaxin. Biochem Biophys Res Commun 1977; 74:1501-1504.

5. Schwabe C, McDonald JK, Steinetz BG. Primary structure of the B-chain of porcine relaxin. Biochem Biophys Res Commun 1977; 75:503-510.

6. Walsh JR, Niall HD. Use of an octadecylsilica purification method minimizes proteolysis during isolation of porcine and rat relaxin. Endocrinology 1980; 107:1258-1260.

7. Büllesbach EE, Schwabe C. Naturally occurring porcine relaxins and large-scale preparation of the B29 hormone. Biochemistry 1985; 24:7717-7722.

8. Schwabe C, McDonald JK. Relaxin: a disulfide homolog of insulin. Science 1977; 197:914-915.

9. Stults JT, Bourell JH, Canova-Davis E et al. Structural characterization by mass spectrometry of native and recombinant human relaxin. Biomed Environ Mass Spectrom 1990; 19:655-664.

10. Haley J, Hudson P, Scanlon D et al. Porcine relaxin: Molecular cloning and cDNA structure. DNA 1982; 1:155-162.

11. O'Byrne EM, Tabachnick M, Anderson LL et al. Characterization of the circulating form of porcine relaxin: Biological activity and terminal amino acids. Endocrinology 1989; 124:2920-2927.

CHAPTER 6

The Initial Approach to Receptor-Binding Site Studies

Clearly, the receptor-interaction site is the most interesting feature of a hormone, but how would one start to look for it? We knew that at the same time several other research groups were having this discussion, among them one of the most successful biotechnology companies, and that gave our efforts a certain sense of urgency.

Step one was to build the porcine relaxin sequence into the known three-dimensional structure of insulin. This was done by two different groups at the time when the porcine relaxin sequence was the only relaxin structure available.[1,2] Consequently, the active site of relaxin was discussed only on the basis of the "parent hormone", insulin. Although suggestions about the active site of relaxin had to be taken with great caution, other interesting predictions were retrieved from these models. It could be demonstrated that porcine relaxin fits into the insulin coordinates without strain and that the surface of the two hormones is so different that relaxin and insulin will not crossreact to the corresponding receptors and antibodies. Unlike insulin relaxin will not bind zinc and, consequently, aggregation properties will be different. These relaxin models remained with us for about 14 years until the X-ray structure of human relaxin became available.[3] As a model insulin was helpful and, although not perfect, it served as structural basis in many of the interpretations shown below. Some of these early models are still in the Brookhaven protein database and can be retrieved by computer modeling applications (pdb files 1rlx, 2rlx, 3rlx, and 4rlx).

As other primary structures of relaxin became available the receptor-binding site could be narrowed down experimentally. It is a justified assumption that when homologous hormones from different species cause a certain effect in a defined biological system that

Relaxin and the Fine Structure of Proteins, by Christian Schwabe and Erika E. Büllesbach. © 1998 Springer-Verlag and R.G. Landes Company.

at least the structure of the receptor-binding site on these hormones must be conserved. It was the next logical step to look at all the available relaxin structures together, i.e., all the A chains discussed in the previous chapter, as aligned by their common feature, the cysteines, and all the B chains likewise (Figs. 4.1, 4.2, 4.3). But structures became available slowly in particular since high structural variability reduced the speed of discovery.

The highly variable A chains nevertheless had cysteines in the same positions (mouse relaxin was not known at this time) and all had a glycine in the A chain loop (A14 in human) (the only exception is the recently discovered marsupial relaxin). Another common feature is the high density of basic amino acid which causes the isoelectric point (PI) of all relaxins to be in the range of pH 8.5-11.0. Besides the general structure there was no specific feature represented in all relaxin A chains such as to make a case for the receptor-interaction site. The constantly appearing glycine in position A14 seemed too small to play a prominent role in receptor-binding.

The B chain alignment gives the same impression except for two arginines in the center helix. These residues were present in every relaxin in the same relative position exactly four residues apart, i.e., in an α-helix they would point into the surrounding water. This feature was uniform enough in all relaxins to qualify for a binding site except that arginines are not the first amino acids that come to mind when one is searching for a receptor-binding region. Regardless, these arginine residues in the B chain became prime suspects and with that realization it became also clear that we needed a total synthesis, a chemical synthesis that would allow us to produce relatively small amounts of a larger variety of derivatives. This meant three years of developmental work which could be used to further sharpen our focus on the ultimate goal, the receptor-binding site.

The first step in delineating a binding-site on a small protein may be the modification of specific residues with equally specific reagents. Porcine relaxin had two tryptophan residues, one in the hydrophobic core of the B chain and one in the C-terminal region of the B chain. These tryptophans could be modified rapidly and specifically with N-bromosuccinimide (Figs. 6.1 and 6.2).[4]

A salt-free sample of relaxin (~0.5 mg = 1 absorbance unit at 282 nm) is dissolved in exactly 1 ml of acidic acid/NaOH buffer (0.2 M, pH 4.7). The solution is placed in a 1 cm quartz cuvette and transferred to a suitable recording spectrophotometer to obtain an initial

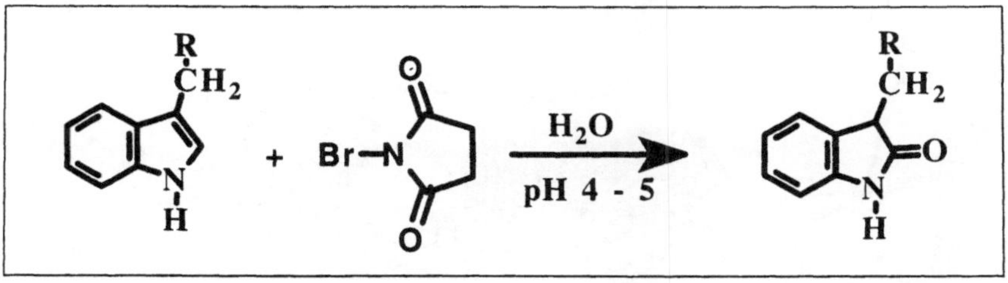

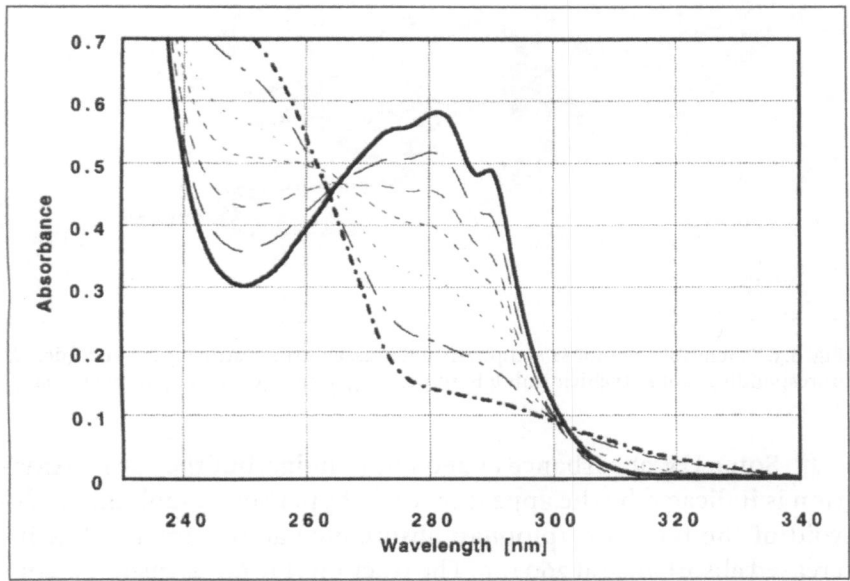

Fig. 6.1. Reaction of an indol ring system with N-bromosuccinimide resulting in oxoindol as the final product.

Fig. 6.2. Reaction of porcine relaxin (0.3 mg/ml in 0.2 M sodium acetate/acetic acid buffer pH 4.7) with N-bromosuccinimide. The UV spectrum of porcine relaxin was recorded (heavy solid line) and N-bromosuccinimide (10 mM) was added in increments of 2 µl and the spectrum recorded. The reaction is complete when no change in UV absorbance is observed (heavy dotted/dashed line). For clarity only every second UV spectrum is shown.

spectrum between 350 and 230 nm. Thereafter the first sample is removed and stored for a later bioassay. The oxidation of tryptophan is accomplished by adding to the photometer cell 2 µl aliquots of a 10 mM solution of N-bromosuccinimide (Fig 6.2). In these experiments samples for bioassay were withdrawn when the absorption at 280 nm had decreased 25, 80, and 100%.

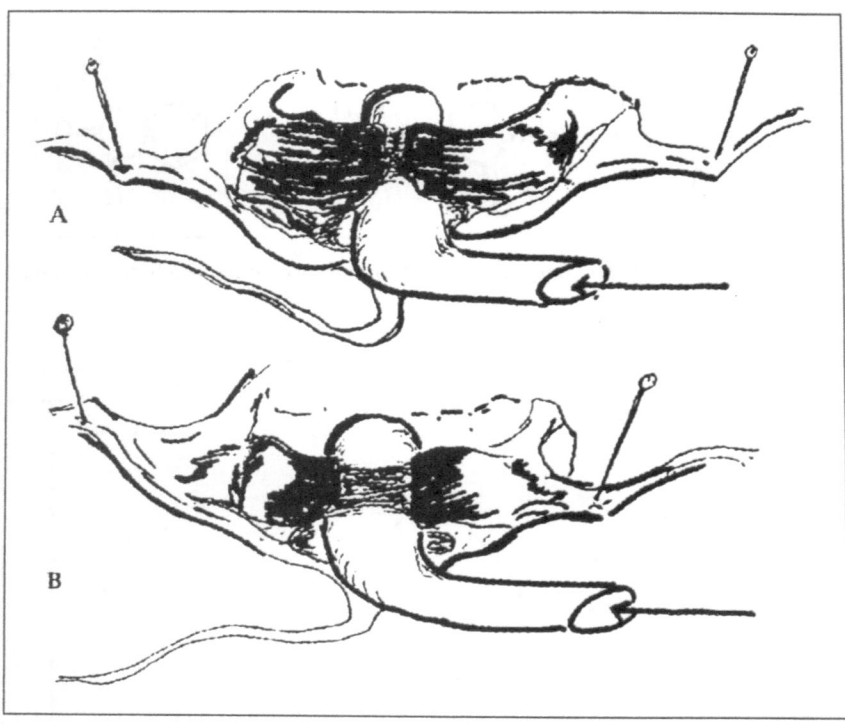

Fig. 6.3. Scheme of the mouse symphysis pubis assay using estrogen-primed mice. A: unresponding mouse (vehicle only); B: responding mouse (0.5 µg of porcine relaxin).

Some UV absorbance at 280 nm remains, but the 100% oxidation is indicated by the appearance of a hyperbolic graph that is devoid of the typical tryptophan absorbance at 295 nm, but has increased absorbance at 260 nm. The reaction (Fig 6.1) is instantaneous and complete and occurs without damage to the primary structure of relaxin. Remaining activity was measured in the mouse bioassay (Fig 6.3).

The samples withdrawn during the oxidation procedure were used to determine the bioactivity remaining as a function of the degree of oxidation. For this purpose young virgin female mice were ovariectomized and injected with 5 µg of a depot estradiol (cypriate or benzoate) after a 3-day recovery period. After five days of exposure to estradiol the mice received an injection of 1 µg of relaxin in the various oxidation states which had been suspended in 100 µl of benzopurpurin-4B (1% solution) as adjuvant. Control groups received benzopurpurin only. The effect of relaxin causes symphyseal widening within the next 8-16 h. After 16 h mice were killed in CO_2

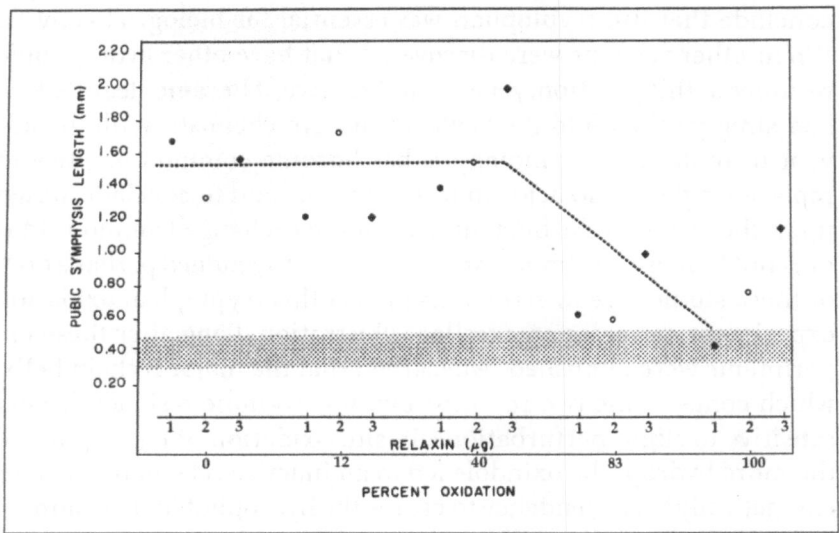

Fig. 6.4. The activity of relaxin in the mouse symphysis pubis assay after varying degrees of tryptophan oxidation by N-bromosuccinimide. At each level of oxidation three groups of mice were injected with the treated relaxin (i.e., 1, 2 and 3 µg/mouse) the shaded bar represents the mean untreated control value ± S.E.M. closed circles, 1 µg/mouse, open circles, 2 µg/mouse and closed diamonds, 3 µg/mouse. mean value for n=10. Reprinted with permission from: Schwabe C, Braddon SA. Biochem Biophys Res Commun 1976; 68:1126-1132.

and the symphyseal joint dissected free of adhering connective tissue and examined under a dissecting microscope that was equipped with a millimeter ocular. The symphyseal joint was transilluminated with fiberoptics and the distance between the shadows of the two pelvic bones was measured (Fig 6.3).

A plot of pubic symphysis length as a function of percent oxidation revealed that one tryptophan residue could be oxidized without affecting biological activity and that the bioactivity disappeared proportional to the conversion of the second tryptophan residue to the oxindole (Fig 6.4).

At the time of the experiment it seemed that one of the two tryptophans was essential for biological activity; a good experiment, a clean observation, but the conclusion was incorrect. Most likely the C-terminal B chain tryptophan, which is very accessible, is oxidized before tryptophan B19, located in a deep hydrophobic pocket in the core of the molecule. Since oxidation of that tryptophan correlated with the loss of biological activity, it was reasonable to conclude that B19 tryptophan was essential for biological activity. When other relaxins were discovered that have other hydrophobic residues at this position, yet were fully active, it became clear that we had simply sailed into the Scylla of protein chemists. With the advent of molecular technology is has become common practice to replace single amino acids in large proteins and to conclude, based upon the effect on the bioactivity, that the exchanged amino acid is or is not biologically important. Conservatively judged perhaps 90% of these studies are in error. Relaxin, via this tryptophan oxidation experiment, provides an excellent illustration. Long after these experiments were published[4] we learned that the major B chain helix, which contains the two receptor-binding arginine residues, is very sensitive to slight perturbations. In situ oxidation of tryptophan to the more hydrophilic oxindole led to an inactive relaxin because its energetically based tendency to escape the hydrophobic environment caused helix distortion and not because tryptophan is important per se. This is a very important message to students of protein-structure function relations.

An other feature shared by all relaxins is a high isoelectric point due to an excess of basic amino acids, arginine and lysine. Even though the lysines did not always occur in the same relative position they were usually clustered in the A chain and occasionally exchanged for arginines. Sequences such as Lys-Lys, Lys-Arg, Arg-Lys, or Arg-Arg were observed. There are several ways to selectively modify primary amino groups such as the lysine side chains and the α-amino group. For a quick survey of the importance of these groups relaxin was exposed to a large excess of succinic anhydride (Fig 6.5).[4] Thus treated relaxin acquires a negative isoelectric point and is devoid of primary amine functions. The arginines still remain unchanged. Fully succinylated relaxin was remarkably active in the biological assay as compared to the unmodified material. Not only did this suggest that primary amino groups are not involved in receptor-binding but also that the receptor-binding site would not be in the A chain because the large substituents on the lysine side chains neighboring the arginine would still hinder receptor-interaction at that site.

The inability to inhibit the relaxin activity by such large-scale modifications caused apprehension. Insulin could be inactivated by acylating the primary amino group of the A chain whereas such a minor manipulation did not disturb relaxin activity. That left only the arginines as the consistently appearing functional group and it

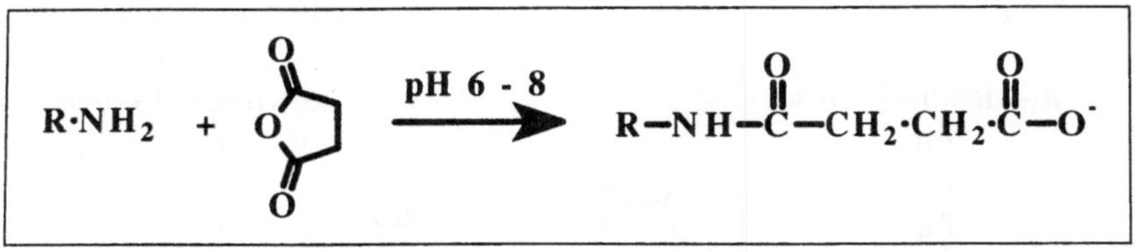

$$R \cdot NH_2 \quad + \quad \underset{O}{\overset{O}{\bigcirc}} \quad \xrightarrow{\text{pH } 6 - 8} \quad R-NH-\overset{O}{\overset{\|}{C}}-CH_2 \cdot CH_2 \cdot \overset{O}{\overset{\|}{C}}-O^-$$

Fig. 6.5. Chemical reaction of a primary amino group with succinic anhydride. Upon opening the cyclic anhydride the amino group forms a peptide bond, leaves the second acidic group to form a negative charge.

was decided to do some preliminary experiments with 1,2-cyclohexanedione (CHD) which reacts with arginines preferentially to form a fairly stable complex without eliminating the positive charge (Fig 6.6). CHD-relaxin proved to be essentially inactive and, although one could not be totally sure, this result strongly implicated the B-chain arginines.[5] One should remember that the previous experiment with succinic anhydride essentially precluded the possibility of receptor interaction for one of the A chain arginines because in porcine relaxin it is situated right next to one of the lysines which could be modified with succinic anhydride without loss of activity; the steric interference alone would make interaction at this A chain arginine impossible.

One might also recall for a moment the words of caution concerning the interpretation of loss of activity after a single amino acid substitution. In contrast to the uncertainty associated with the loss of activity, retention of activity after modification by either amino acid replacement or side chain modifications is subject to unambiguous interpretation.

In case of the arginines, modification with 1,2-cyclohexanedione (CHD) could be narrowed down by successive exclusion of other arginine residues present in the molecule. Even though CHD modifies all arginines on the surface, and may occasionally modify a lysine, considering the lack of effect of the succinic anhydride modification, the inactivation of relaxin by CHD could be ascribed to the two B chain arginines in positions 12 and 16 porcine with some degree of confidence. The arginine occurring in position B29 cannot be important because a B28 relaxin is fully active. The only question remaining concerned the relative bulk effect of CHD. By capping the B chain arginines 13 and 17 with CHD, did we prevent the molecule from simply achieving the proper juxtaposition to the receptor and

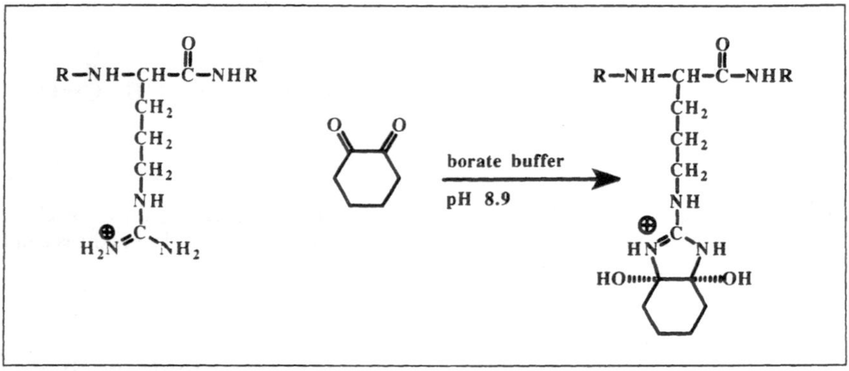

$$R-NH-CH-\overset{\overset{O}{\|}}{C}-NHR$$

$$CH_2$$
$$CH_2$$
$$CH_2$$
$$NH$$
$$H_2N-\overset{\oplus}{C}-NH_2$$

borate buffer
pH 8.9

$$R-NH-CH-\overset{\overset{O}{\|}}{C}-NHR$$

$$CH_2$$
$$CH_2$$
$$CH_2$$
$$NH$$
$$HN-\overset{\oplus}{C}-NH$$
$$HO\cdots\diagdown\diagup\cdots OH$$

Fig. 6.6. Reaction of the guanidinium group in arginine with 1,2-cyclohexanedione. The resulting modification still retains a positive charge.

thus hindered the true receptor-interacting residues from making contact? The final answer had to wait until completion of the total synthesis of relaxin, which would allow one to make isosteric substitutions. That story, however, will have to wait since our narrative is only in the mid 1980s. The increased incidence and severity of strong language in our laboratory made it quite clear that synthesis was well under way.

Meanwhile a synthesis of porcine relaxin and human relaxin had been reported at a relaxin symposium in 1982 which was, unfortunately, of little use because it was never published in refereed journals and did not contain enough information to show that the right molecules were obtained.[6] In addition, the reported difference in bioactivity between native porcine and synthetic human relaxin is incompatible with the literature on relaxin. Thus it was necessary to invest three years into the design and documentation of a reproducible synthesis which included site-specific sequential disulfide bond formation. History provides good reasons for rigorous documentation which is often taken as less important by our more biologically oriented colleagues. Many years ago the first synthesis of human growth hormone was hastily reported without much documentation. There were claims of bioactivity until it became clear that the synthesis had been made according to an inaccurate primary structure.[7,8] Humans' ability to be concise is limited by natural laws and the way the mind works, but with these provisos care and conscientiousness always take first place.

The point had been reached where nothing "easy" could be done about the active-site determination. Was one or were both of the arginines important and, if one only, which one and what residue could one substitute for arginine without disturbing other parameters? It seemed quite clear that this was the important region in spite of the fact that there was no good precedent for arginines as binding residues, and that several X-ray crystallographers spoke with somewhat lofty affirmatism about the importance of the N-terminal A chain region,[9] in particular the serine and lysine residues, as part of the receptor-binding region. Until these questions could be settled decisively by a synthetic route it was decided to pursue a thorough investigation of other areas of porcine relaxin and to study the effect of other regions on the overall structural integrity of relaxin as compared to our standard, the porcine B29 molecule.

References

1. Bedarkar S, Turnell WG, Blundell TL et al. Relaxin has conformational homology with insulin. Nature 1977; 270:449-451.
2. Isaacs N, James R, Niall H et al. Relaxin and its structural relationship to insulin. Nature 1978; 271:278-281.
3. Eigenbrot C, Randal M, Quan C et al. X-ray structure of human relaxin at 1.5 Å: Comparison to insulin and implications for receptor binding determinants. J Mol Biol 1991; 221:15-21.
4. Schwabe C, Braddon SA. Evidence for one essential tryptophan residue at the active site of relaxin. Biochem Biophys Res Commun 1976; 68:1126-1132.
5. Büllesbach EE, Schwabe C. On the receptor binding site of relaxin. Int J Peptide Protein Res 1988; 32:361-367.
6. Tregear GW, Du YC, Wang KZ et al. The chemical synthesis of relaxin. In: Bigazzi M, Greenwood FC, Gasparri F, eds. Biology of Relaxin and Its Role in the Human. Excerpta Medica, 1983:42-55.
7. Li CH, Yamashiro D. The synthesis of a protein possessing growth-promoting and lactogenic activities. J Am Chem Soc 1970; 92: 7608-7609.
8. Li CH, Dixon JS. Human pituitary hormone XXXII the primary structure of the hormone: Revision. Arch Biochem Biophys 1971; 146:233-236.
9. Blundell T, Gowan LK, Schwabe C. Relaxin—a member of the insulin family? In: Bigazzi M, Greenwood FC, Gasparri F, eds. Biology of Relaxin and Its Role in the Human. Excerpta Medica, 1983:14-21.

CHAPTER 7

The N-Terminal Region of the Relaxin A Chain

R elaxin as referred to in this chapter is B29 porcine relaxin which has an A chain starting with arginine and ending with cysteine, and a B chain starting with pyroglutamic acid and ending with arginine in position B29 (hence B29 relaxin). Due to the presence of pyroglutamic acid in position B1, the N-terminal region of the A chain lends itself to modifications or even to total exchange. The questions to be answered then were: 1) is the A chain N-terminal region important, and if so, 2) what is its role?

Removal of the N-terminal end of the A chain by preparative Edmann degradation required protection of all primary amine side chains which would otherwise be irreversibly converted to the phenylthiourea during Edman degradation. This could be achieved either by the reversible reaction with citraconic anhydride, followed by other steps, or by the direct reaction with methylsulfonylethyl-oxycarbonyl-(Msc)-N-hydroxysuccinimide ester (MscONSu). The summary in Figure 7.1 shows the reaction in a weakly buffered solution at neutral pH with citraconic anhydride which reacts with all primary amino groups. The weak buffer is soon overwhelmed by the citraconic acid which appears as a byproduct so that the pH drops to approximately 4.5. At that pH the ε-amino groups of lysines lose the citraconyl group whereas the α-amino group modification remains stable. The free ε-amino groups of lysines can now be modified with MscONSu (in which Msc is the protecting group and ONSu is the activating group that yields N-hydroxysuccinimide as leaving product. It is therefore wrong to speak of succinylation in those cases as it is seen occasionally in the literature). The MscONSu groups will rapidly react with the free ε-amino groups of the lysines. The citraconyl group will be removed from the α-amino group of the A

Relaxin and the Fine Structure of Proteins, by Christian Schwabe and Erika E. Büllesbach. © 1998 Springer-Verlag and R.G. Landes Company.

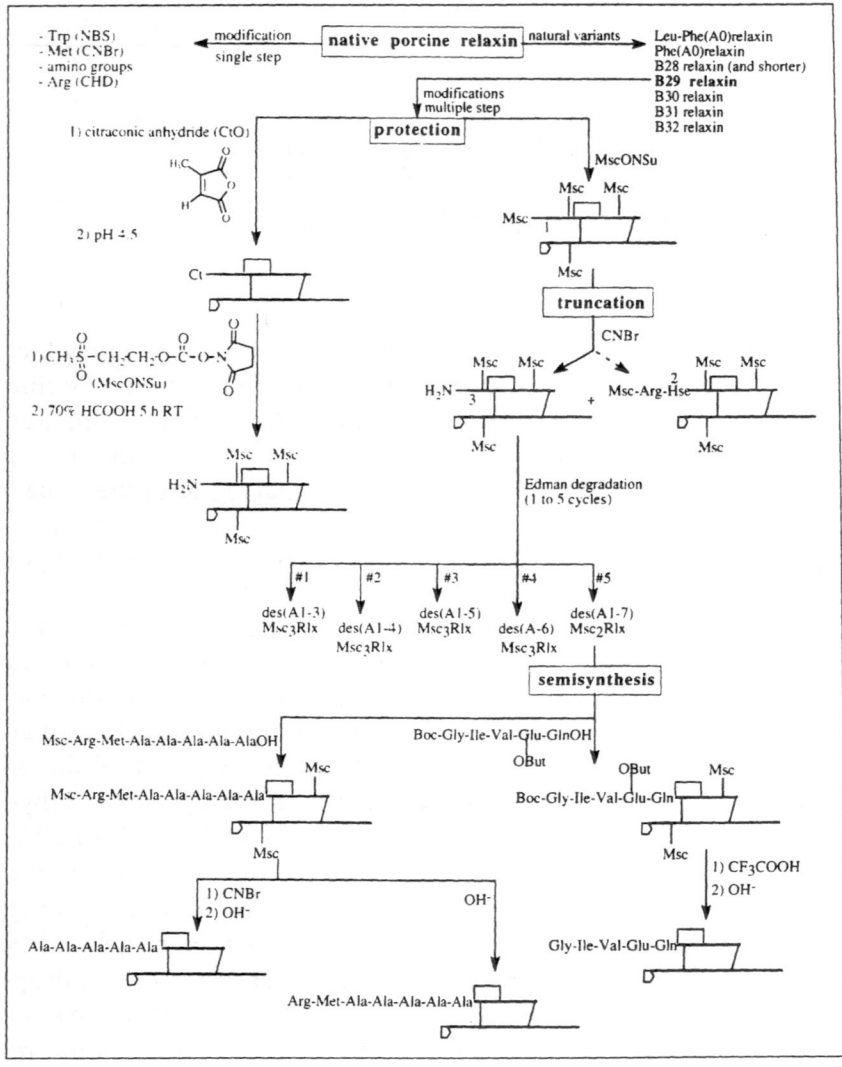

Fig 7.1. Scheme of chemical modifications and semisyntheses performed with B29 porcine relaxin. (Abbreviations: Ct, citraconyl; Hse, homoserine; Msc, methylsulfonylethyloxycarbonyl; NBS, N-bromosuccinimide; OBut, tert butyl ester; Rlx, relaxin)

chain by subsequent exposure to 70% formic acid for 5 hr. The end product is a side chain-protected relaxin with a free α-amino group.[1] This molecule is now ready for sequential preparative Edman degradation whereby seven residues can be removed without permanently modifying the rest of the molecule.

The second scheme calls for the direct reaction with MscONSu and takes advantage of the presence of a cyanogen bromide (CNBr)-susceptible methionine in position A2. Exposure to CNBr in 70% trifluoroacetic acid will give two products one of which has lost the dipeptide Arg-Met and begins with Thr and one (the other 20% of the relaxin) is not fragmented but rather shows conversion of methionine to homoserine.[2] The citraconyl derivative provided us with des-arginine relaxin after the first step of degradation whereas the direct Msc-protection method, followed by CNBr fractionation, began with Thr A3. At every stage of degradation a sample of relaxin was removed from the reaction medium for biological assay in mice. While the loss of arginine and methionine caused no change in bioactivity the loss of further amino acids successively reduced biological activity, and porcine relaxin was totally inactivated by the removal of four or more residues from the N-terminal end (Fig. 7.2).[2] Could one restore the bioactivity by other helix-forming peptides such as the N-terminal region of insulin or even an oligo-alanine helix? First the N-terminal peptide of porcine insulin was synthesized in its protected form as shown in Figure 7.3 and, when this material after coupling to relaxin and deprotection proved to be nearly as active as B29 relaxin,[3] we decided somewhat mischievously to put a penta-alanine onto the N-terminal end of the A chain to see whether relaxin would notice. As simple as a penta-alanine may seem to the reader it turned out to be impossible to make and to couple to relaxin because of its utter insolubility. Apparently alanine peptides of a certain length produce β-pleated sheets rather than helices and become what is lovingly referred to as a "stone" among peptide chemists. It sits on the bottom of a tube and will do nothing. An alternative route was to produce this peptide with an Arg-Met at the end in order to increase solubility and then to couple Arg-Met penta-alanine onto the shortened A chain of Msc_2relaxin. This worked out great in every respect from the solution of the technical problems to the fact that two different derivatives had been obtained. The Arg-Met penta-alanine relaxin could easily be converted into the penta-alanine relaxin by CNBr treatment.[3]

All of these derivatives were deprotected, purified, and assessed again by the mouse pubic ligament assay which is the only truly reliable indicator of relaxin activity. The results of these assays as depicted in Figure 7.2 clearly show deactivation as a function of removal of N-terminal A chain residues and the reversal of this effect

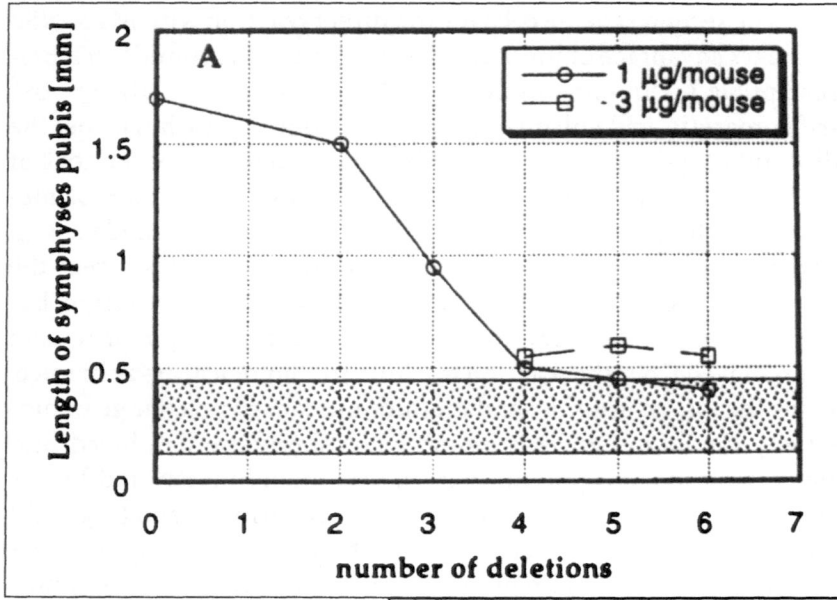

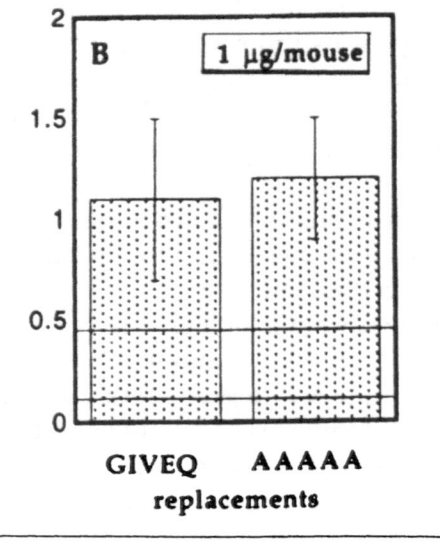

Fig 7.2. Mouse symphysis pubis assays. (A) Activity of porcine relaxins truncated at the N terminus of the A chain. Two series of experiments were performed; in the first all relaxin analogs were administered at a dose of 1 μg/mouse. Analogs truncated by four and more amino acid residues appeared inactive and those were given at a dose of 3 μg/mouse in a second experiment. The response is displayed as function of truncation. (B) Mouse symphysis pubis assay of porcine relaxin analogs with replacements of the A chain helix by the insulin segment (GIVEQ) and by penta-alanine (AAAAA). Both analogs were given at a dose of 1 μg/mouse.

by the addition of the insulin peptide as well as the penta-alanine or the Arg-Met-penta-alanine. After many manipulations about 50% of the activity of unmodified and untreated porcine relaxin was recovered but, remarkably, there was no significant difference in bioactivity between relaxins with either the insulin segment or the penta-alanine substitution. Although the CD spectroscopy showed slight

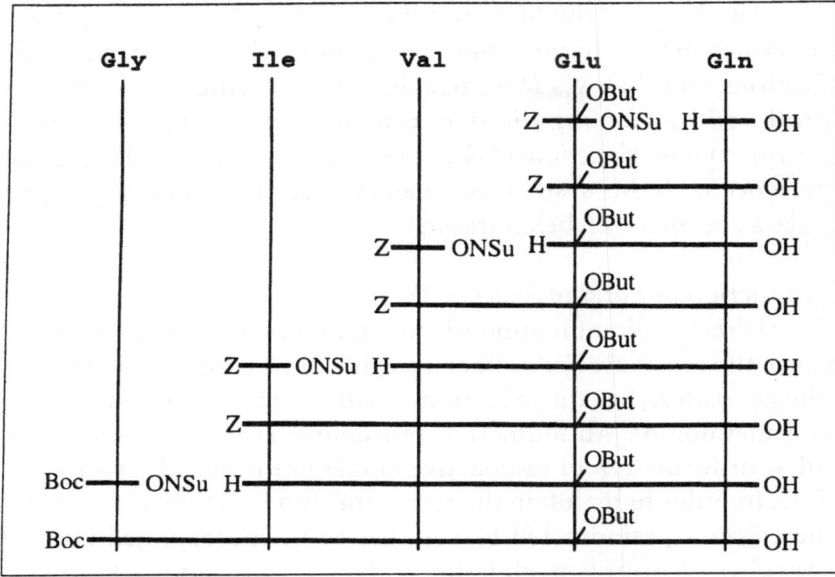

Fig 7.3. Chemical synthesis of the N-terminal peptide of the insulin A chain (according to Föhles, dissertation RWTH Aachen, 1972). The following protecting groups are used: amino protecting groups Z = carbobenzoxy group reversible by catalytic hydrogenation; Boc = tert.butoxycarbonyl labile toward trifluoroacetic acid; carboxyl protecting group: OBut = tert butyl ester, labile in trifluoroacetic acid. Carboxyl activation was achieved by ONSu = N-hydroxysuccinimide ester. Other abbreviations are, H = free amino group, OH = free carboxyl group.

changes at 220 nm, the relaxin hybrids had retained their high helix content and therefore we felt entitled to conclude that the A chain N-terminal region has no other function than to provide an α-helix that would help to maintain the integrity of the structure by restricting water access to the hydrophobic core. These experiments encouraged the prediction that helix-breaking amino acids should not occur in a relaxin A chain closer than 5 residues to the first cysteine. Clearly, this region had nothing to do with the receptor-binding site and that result was compatible with the CHD study which implicated the two B chain arginines. Relaxin appeared to be writing a textbook on protein design by giving credence to the generally accepted statement that much of the excess structure in any biologically active protein outside the catalytic or binding site may function to keep the molecule in a proper conformation. This was certainly more than one could expect a priori to learn from a "minimal" (referring to its smallness) protein.

The fact that the Msc$_4$-relaxin still showed biological activity suggested that the lysine residues could be used for functional modifications and that, via Msc$_3$-relaxin, a site-specific tracer could be produced by reacting the free α-amino group with a prelabeled 4-hydroxyphenylacetic acid N-hydroxysuccinimide ester. Further experience with this system led to an even simpler approach to produce a specifically labeled tracer.

Production of Porcine Relaxin Tracer[4]

Direct radioiodination of porcine relaxin is not possible because the molecule has neither tyrosine nor histidine and, to complicate matters, has the oxidation-sensitive amino acids tryptophan and methionine. All iodination procedures require an oxidant for the transformation of radioactive iodide to the volatile reactive iodine. In order to maintain the structural and functional integrity of the tracer a preiodinated reagent has to be incorporated into the protein. Although prelabeled ^{125}I-[3-(3,5-diiodo-4-hydroxy-phenyl)propionic acid N-hydroxysucciniimide ester (Bolton Hunter reagent) is commercially available it is expensive and subject to slow hydrolysis. Furthermore it is easily and better made in situ. One microliter of a freshly prepared stock solution of the N-hydroxysuccinimide ester of 3-(4-hydroxyphenyl)propionic acid (= desamino tyrosine) (0.26 µg/µl) in waterfree, basefree N,N-dimethylformamide (DMF) is placed into the tip of a 500 µl Eppendorf vial. The vial is kept on ice and 5 µl of 0.25 M phosphate buffer pH 7.5 is placed on the wall of the vial such that the two do not mix. To this droplet of buffer 10 µl = 5 mCi Na^{125}I (Amersham, IMS.300) is added and the mixture vortexed. In order to avoid hydrolysis of the N-hydroxysuccinimide ester the following additions are made as quickly as possible and in the listed order: 5 µl of chloramine T (2 mg/ml in 0.25 M phosphate buffer pH 7.5), 10 µl of sodium thiosulfate (x5 water) (50 mg/ml in 0.25 M phosphate buffer pH 7.5), 10 µl of KI (20 mg/ml in 0.25 M phosphate buffer pH 7.5), and 10 µl of porcine relaxin (1 mg/ml in 0.25 M phosphate buffer pH 7.5). The coupling reaction is then performed for 3 h at room temperature, the reaction mixture loaded onto an HPLC system and separated. Each peak on the chromatogram subsequent to the large peak of unreacted relaxin represents a different substitution locus or mono vs. disubstitution (Fig. 7.4).

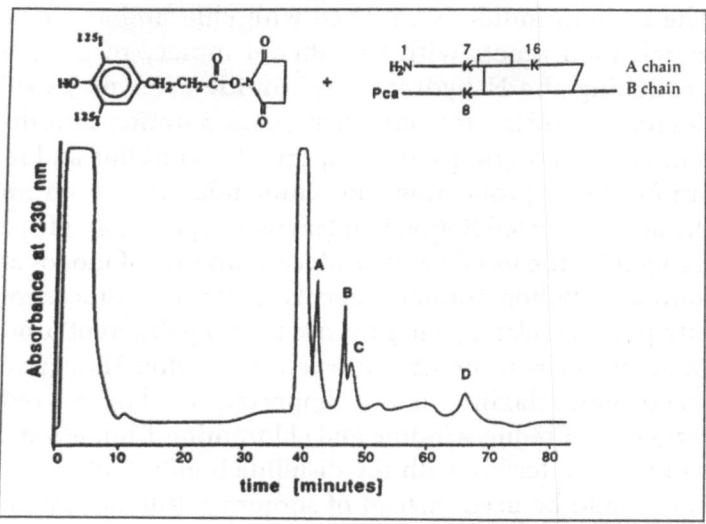

Fig 7.4. HPLC elution profile of porcine relaxin tracer. Upper panel: reactants indicating the four potential reactive sites of porcine relaxin. Lower panel: the radioactive peaks correlate to the different monosubstituted relaxin: A = A7, B = A1, C = A16, D = B8. The main UV absorbing peak is identical to unsubstituted porcine relaxin. HPLC conditions: Aquapore C_8 (2.1 x 30 mm). Solvent A consisted of 0.1% trifluoroacetic acid in water and solvent B of 0.1% trifluoroacetic acid in 80% acetonitrile/20% water. Elution was achieved with a linear gradient from 25% B to 45% B in 90 min at a flow rate of 100 μl/min. UV absorbance was determined at 230 nm and fractions were collected by hand into 100 μl of 1% bovine serum albumin and radioactivity determined. Reprinted with permission from: Yang S, Rembiesa B, Büllesbach EE et al. Demonstration of ligand binding in symphyseal tissues and uterine membrane. Endocrinology 1992; 130:179-185. © The Endocrine Society.

There are few tricks to this reaction:

1. In order to avoid hydrolysis of the active ester the radioactive iodide needs to have a low pH (usually pH 8-9). Mixing of the iodide with the N-hydroxysuccinimide occurs in a strong phosphate buffer at pH 7.5. Once the active ester is in contact with water hydrolysis will occur, and when the ester is hydrolyzed the final labeling of relaxin will not take place. Therefore from the moment the active ester comes in contact with water all subsequent steps have to be fast but not hasty. Under normal circumstances adding and mixing of the components may not take longer than 60 seconds.

2. The sodium iodide is oxidized with chloramine T to iodine which then reacts with the phenol moiety of the reagent, producing the N-hydroxysuccinimide ester of 3,5-diiodo-desaminotyrosine for immediate incorporation into the protein via amino groups. It is important that iodine and reagent are in proper proportions. Recommended is to use 1 nmol of the non-iodinated Bolton Hunter reagent per 2.5 nmol of $Na^{125}I$. Using too little iodide will produce a mixture of mono- and di-iodinated Bolton Hunter reagents which, upon incorporation into porcine relaxin, will give rise to many different products.

3. Oxidant needs to be absent when the Bolton Hunter reagent reacts with relaxin. It is very important to choose a reducing reagent that reduces iodine and chloramine T but is not strong enough to interfere with the disulfide bonds. Sodium thiosulfate should be used instead of sodium sulfite or cysteine. Sodium thiosulfate can be used in excess. It is good practice to perform a preliminary test to show that all reagents are in working condition. If, after the addition of the recommended concentrations and volumes of chloramine T and 10 µl of the KI solution, a yellow color appears which disappears again after the addition of 10 µl of thiosulfate, then all concentrations are correct.

4. The condensation of Bolton Hunter reagent with relaxin is a bimolecular reaction and concentration-dependent. For these reasons we always use the iodide in concentrations as high as possible. Reactions with more dilute radioactive iodine have resulted in very low condensation yields and therefore small quantities of tracer.

By the same method one could label relaxin with biotin. The introduction of biotin was studied in more detail than the reaction with radioactive Bolton Hunter reagent. At that time it was learned that one cannot direct the position of modification by pH or solvent techniques that were very useful during the semisynthesis of insulin-analogs.[5,6] Reactions with relaxin always give rise to mixtures with different product distributions.[7] This method has been used by Sherwood for in vivo detection of biotin-labeled relaxin in rat tissues.[8,9]

References

1. Büllesbach EE, Schwabe C. Preparation and properties of alpha and epsilon-amino-protected porcine relaxin derivatives. Biochemistry 1985; 24:7722-7728.
2. Büllesbach EE, Schwabe C. Preparation and properties of porcine relaxin derivatives shortened at the amino terminus of the A chain. Biochemistry 1986; 25:5998-6004.
3. Büllesbach EE, Schwabe C. Relaxin structure: Quasi allosteric effect of the NH2-terminal A chain helix. J Biol Chem 1987; 262:12496-12501.
4. Yang S, Rembiesa B, Büllesbach EE et al. Relaxin receptors in mice: Demonstration of ligand-binding in symphyseal tissues and uterine membrane. Endocrinology 1992; 130:179-185.
5. Geiger R, Geisen K, Summ HD et al. [A1-D-alanine]insulin. Hoppe Seyler's Z Physiol Chem 1975; 356:1635-1649.
6. Geiger R, Langner D. Insulin analogs with B-chains shortened at the N-terminal end. Selective Edman degradation of the insulin B chain. Hoppe-Seyler's Physiol Chem 1973; 354:1285-1290.
7. Büllesbach EE, Schwabe C. Monobiotinylated relaxins: Preparation and chemical properties of the mono(biotinyl-ε-aminohexanoyl) porcine relaxin. Int J Peptide Protein Res 1990; 35:416-423.
8. Min G, Sherwood OD. Identification of specific relaxin-binding cells in the cervix, mammary glands, nipples, small intestine, and skin of pregnant pigs. Biol Reprod 1996; 55:1243-1252.
9. Kuenzi MJ, Sherwood OD. Immunohistochemical localization of specific relaxin-binding cells in the cervix, mammary glands, and nipples of pregnant rats. Endocrinology 1995; 136:1367-1373.

CHAPTER 8

The Total Synthesis
of Human Relaxin

Preliminary work on relaxin, or rather on the relaxin perimeters, had built up such a "curiosity pressure" that the trials and tribulations of a total synthesis became the lesser problem. When a peptide chemist faces a new synthesis he will immediately divide the problem into two major categories, one the overall plan of development of a primary structure (fragment condensation versus synthesis by sequential addition of amino acids) and secondly the decision concerning semipermanent side chain protections that must be compatible with the overall plan. If, for example, repeated exposure to weak acid is part of the synthesis of the primary backbone, acid labile semipermanent protecting groups can only be used if they require a much stronger acid for cleavage (trifluoroacetic acid vs. hydrogen fluoride) and so on. We will describe the synthesis of relaxins, including the special problems, in detail so that a student may derive sufficient information from the general concept for the synthesis of other peptides under similar conditions. For that reason a few of the approaches that did not work will be useful. In chapter 9 the problem of proper documentation of syntheses will be discussed as well as the recording of this type of research for it to become part of the serious scientific literature.

The uninitiated will describe peptide synthesis as the chemical formation of a polypeptide with a specific sequence. This is absolutely true simply because the statement says absolutely nothing. The chemist deals with differential protection, carboxyl group activation and specific condensation to amino groups (Fig. 8.1) and this is where success and failure are determined. Why protection in the first place; it makes the process more tedious, and furthermore, biosynthesis

Relaxin and the Fine Structure of Proteins, by Christian Schwabe and Erika E. Büllesbach. © 1998 Springer-Verlag and R.G. Landes Company.

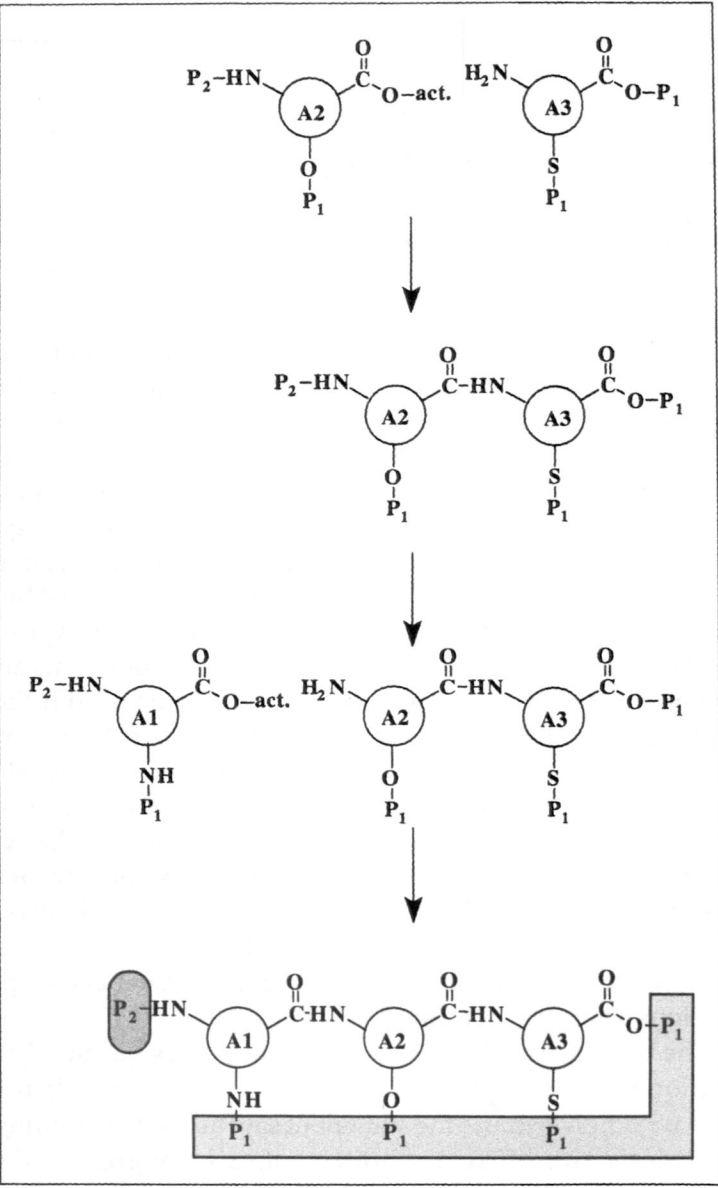

Fig. 8.1. Synthesis of a model tripeptide consisting of three tri-functional amino acids. The side chains and the carboxyl group of residue A3 are protected via semipermanent protecting groups (P_1). The α-amino group of the incoming amino acid A2 and A1 are protected by the temporary protecting group (P_2). The α-carboxyl groups of amino acid A2 and A1 are activated for a nucleophilic addition and elimination reaction. P_2 can be removed selectively while P_1 remains stable.

proceeds without protecting groups. Of course the biological process is ideal and several attempts have been made to mimic it in vitro,[1] but chemistry in our hands cannot repeat the incredible feats that enzyme-catalyzed reactions can achieve in terms of specificity and reaction conditions. For example, an electrophile in an uncatalyzed chemical reaction will react with all nucleophiles; the product distribution is determined by the stochiometry, electron density and sterical effects. Nucleophile and electrophile in an enzyme-catalyzed reaction will be sterically directed to the place that is compatible with the enzyme-binding site thus providing inimitable specificity under very mild conditions. In a chemical reaction we have to direct the reactive group to the appropriate receiving group by simply covering all other potentially reactive centers by protecting groups. The electrophile, in this case the activated carboxyl group, therefore sees only one nucleophile, and it either reacts there or not at all. Organic solvents are used for these procedures because water is a nucleophile and would compete with the targeted reaction. Solubility is another good reason to use organic solvents and choosing those that have no functional groups that could compete with the peptide chemistry is equally important. For multiple condensation either an amino group or a carboxyl group needs to be liberated from its protecting group before it can react with another amino acid. These temporary protecting groups have different properties than the semipermanent protecting groups which have to be stable during coupling and deprotection reactions and remain on the growing peptide until the desired sequence is assembled.

To demonstrate the synthesis of such a temporarily protected amino acid we have chosen the 9-fluorenymethyloxycarbonyl citrulline (Fmoc-CitOH). The Fmoc group is the most common α-amino protecting group where citrulline is a natural amino acid not incorporated during ribosomal protein biosynthesis (Fig. 8.2).

Synthesis of Fmoc-Citrulline (according to Ten Kortenaar et al[2])

Citrulline (4.35 g, 25 mmol) was dissolved in 25 ml of water in the presence of 25 mmol triethylamine in a 100 ml Erlenmeyer. A pH electrode was placed into the solution and under stirring 7.3 g (24 mmol) Fmoc-ONSu, dissolved in 25 ml of dry acetonitrile, was added at once. The pH was adjusted to pH 8.5 to 9 using triethylamine. After the pH remained constant (about 15 min) the reaction was continued for another 30 min. Thereafter the mixture was stirred into

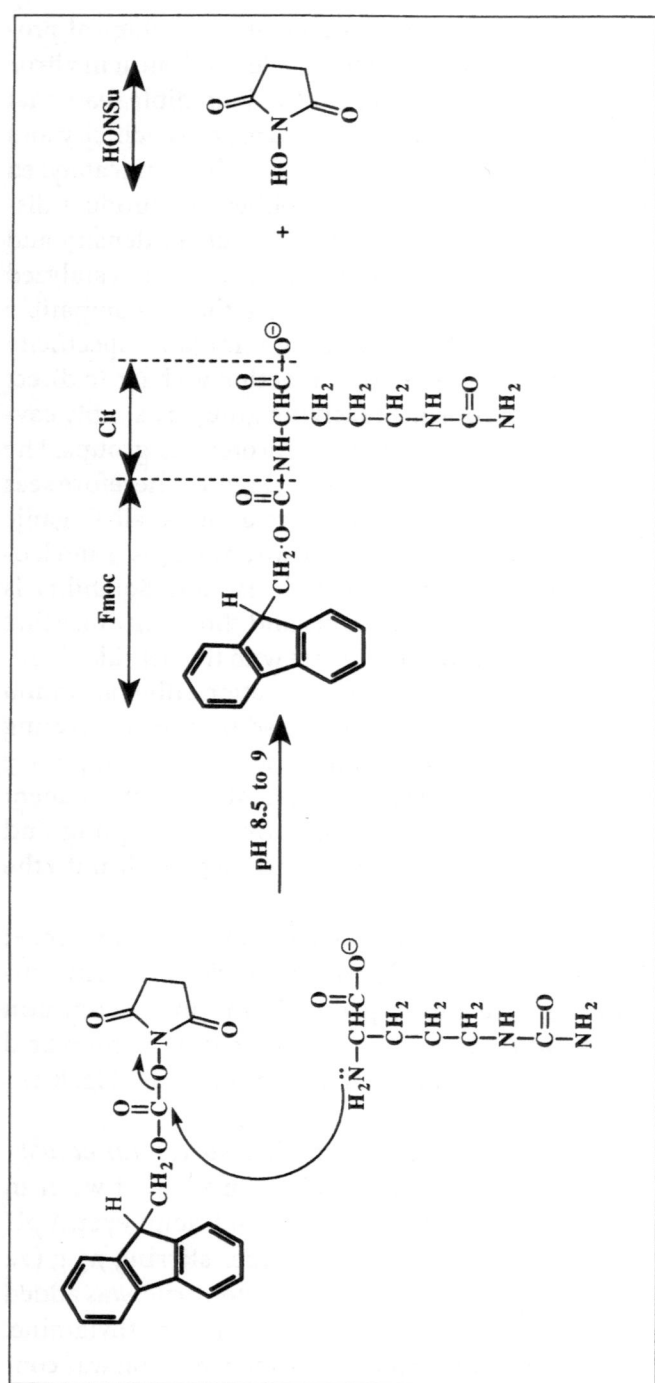

Fig. 8.2. Chemical synthesis of Fmoc-CitOH as an example for amino group protection.

100 ml of 1.5 M HCl. The resulting oil was extracted with ethylacetate (3 x 50 ml), the pooled organic layers washed with 20 ml of water (2x) and then dried over $MgSO_4$. The $MgSO_4$ was filtered off and the ethylacetate removed in vacuo. The resulting oil was crystallized from acetonitrile (m.p. 120-122°C, yield 7.75 g = 78.4%).

Run on t.l.c. on fluorescent silicagel F254 (Merck, Darmstadt, Germany) in the following solvent systems:
1. butanol-1 : acetic acid : water 4/1/1 v/v/v Rf = 0.83;
2. chlorofom : methanol : acetic acid 95/5/3 v/v/v Rf = 0.10
3. in chloroform : methanol : water 70/30/5 v/v/v Rf = 0.25
   The product is ninhydrin-negative but visible as a dark spot on a fluorescing background (quenching).
   tBoc citrulline is made similarly. Now both compounds can be purchased. Stay tuned however for a good reason, to at least be able to make crucial derivatives in your own laboratory.

Thirty-five years after Bruce Merrifield's pioneering work on the automation of the condensation process[3] most peptides are synthesized on fully automatic synthesizers that control progress of the synthesis through a feedback loop and adjust for difficult condensations by increasing the coupling time. Automatic peptide synthesis is begun by coupling the C-terminal amino acid to a solid support and continued by repetitive deprotection of amino groups and addition of further amino acids. This very efficient process is used to generate most of the small peptides needed for biomedical research.[4,5] Beyond automation lies a still open research area, i.e., the generation of larger three-dimensional structures (proteins) and mimetics of peptides and proteins with potential pharmaceutical application. While one is synthesizing two-dimensional arrays, it is the "third dimension" that is assumed by a peptide upon contact with water that produces bioactivity. The total synthesis of relaxin tells an interesting, if not intriguing, story about the relationship of all aspects of protein structure and biological function.

Small two-chain disulfide-linked proteins, such as insulin, relaxin, relaxin-like factor, bombyxin and others, present the problem of proper disulfide bond formation. True, native chains, or better unmodified chains, will take up the native configuration to a certain degree and will combine properly during air oxidation.[6-8] In vivo these hormones are produced as a single chain precursor whereby the connecting peptide between the A and B chain plays the role of a quasi chaperone. The prohormone folding is guided so that the appropriate cysteine residues come together upon oxidation to form the disulfide bridges.[9] The connecting peptide is then removed by endopeptidases which in turn are guided by the presence of target sites, such as an accumulation of two to four basic residues in a row. Further processing may occur by amino- or carboxypeptidases until the stable configuration has been reached.[10,11] Compared to that process the in vitro oxidation and combination of chains is very inefficient. The yield of the proper disulfide-bonded product can go all the way to zero when substitutions are made[7] and one must remember that for structure/function work the non-natural derivatives are most important. There are several ways to get around this problem and all of them are not simple. One could synthesize the three disulfide bonds chemically in a site-directed way, or one could mimic nature by including a connecting peptide that supports folding and subsequent correct oxidation of the six cysteines.

The first approach was to try the chemical synthesis of the three disulfide bonds by segment condensation. Figure 8.3 shows the molecule divided into four synthetic segments, the B chain with two different cysteine protecting groups, and three peptides of the A chain of which the center peptide (A7-17) contains three cysteine residues and the C-terminal peptide one cysteine residue. The underlying principle is to use different protecting groups for the cysteines that can be liberated two at a time. The first disulfide bond can then be formed before the next set of cysteines is liberated or activated. The ideal combination has been tested in detail during the synthesis of the internal disulfide loop in insulin.[12-14] In our example the trityl-protecting groups are removed first and the disulfide bond formed

Fig. 8.3 (opposite). Synthesis of human relaxin by segment condensation using differential protection of the cysteine side chains. For clarity only the cysteines-protecting groups are shown. Segments as outlined in the gray shaded box were synthesized by solid phase methodologies. The B chain was generated with an acetamidomethyl (Acm) group in position B11 and a free sulfhydryl group in position B23. The intrachain disulfide loop of the A chain was formed from trityl (Trt)-protected cysteines in position A11 and A15 while the acetamidomethyl (Acm) protection in A12 remained stable. The C-terminal cysteine of the A chain segment A(18-24) was S-activated and reacted with the free sulfhydryl group of the B chain as shown in Fig. 8.4. The α-carboxyl groups of the A chain segments were preactivated and condensed to a single amino group generated as required at position A18 and A7, respectively (all other amino groups were protected during this process).

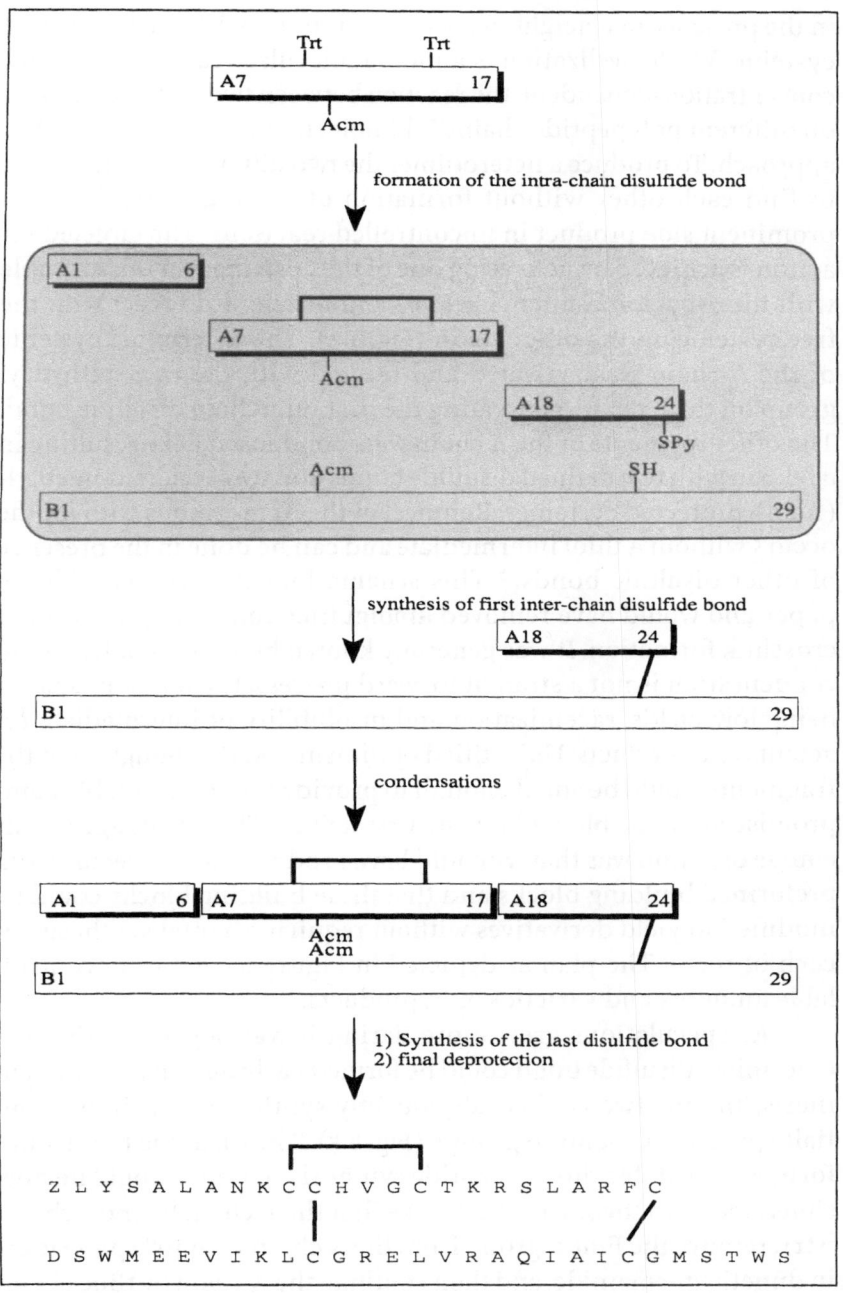

in the presence of a neighboring acetamidomethyl-(Acm)-protected cysteine. While cyclization is a monomolecular reaction that is not concentration-dependent, the reaction between two cysteines located on different polypeptide chains is bimolecular and requires another approach. To produce a heterodimer the two different cysteines have to find each other without formation of homodimers, the most prominent side product in uncontrolled reactions. The targeted reaction is achieved by activating one of the cysteines (in this example with the 2-pyridinesulfenyl group) and forcing it to react with the free cysteine on the other chain (Fig. 8.4). The C-terminal cysteine of the A chain was activated and reacted with the free sulfhydryl group on the B chain, generating the first interchain disulfide bond. The other segments of the A chain were condensed next, resulting in a relaxin with two defined disulfide bonds and two acetamidomethyl-(Acm)-protected cysteines. Removal of the Acm-groups with iodine occurs without a thiol intermediate and can be done in the presence of other disulfide bonds.[15] This schema looked very plausible on paper and would have removed ambiguities concerning the proper crosslink formation. It was generally known however that fragment condensation is not a straight forward process, the major problems being low yields, racemization and insolubility of intermediate reactants and products. Unjustified optimism and the thought that the fragments would be small enough to provide for an acceptable compromise are to be blamed for the first failure. The advantage of this course of action was that we could have produced the molecule from preformed building blocks and that these building blocks could be modified to yield derivatives without requiring a total synthesis for each of them. The plan as depicted in Figure 8.3 led to uncontrollable amounts and varieties of byproducts.

As speculations grew more daring it was suggested that the C-terminal disulfide bond could be formed early during peptide synthesis, and the two chains subsequently synthesized with differentially protected α-amino groups (Fig. 8.5). The disulfide bond, once formed, was stable during peptide synthesis so that it would be possible, at least in theory, to synthesize first the B chain by tBoc-chemistry, remove the Fmoc-group from the A chain with 10% piperidine in dimethylformamide, and then continue the A chain by tBoc chemistry. This approach would reduce the formation of the remaining crosslinks to monomolecular reactions which should be very fast and efficient. Model peptides, somewhat smaller than relaxin, were

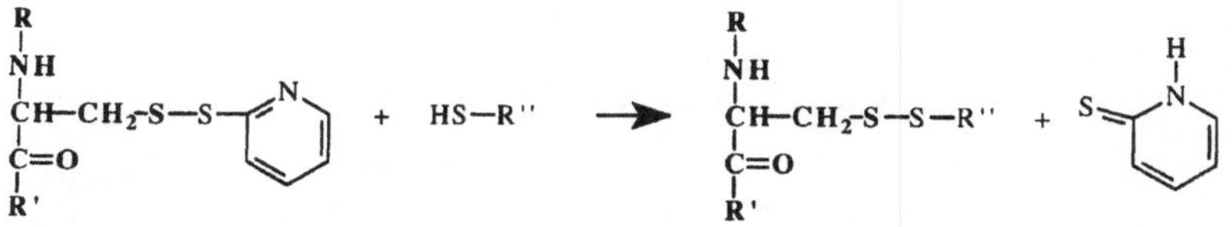

Fig. 8.4. Scheme of the synthesis of an unsymmetrical two-chain disulfide peptide. The first polypeptide contains an activated S-pyridylsulfenyl cysteine which reacts specifically with the free sulfhydryl of the second chain.

produced with encouraging results.[16] A few more residues added in the case of relaxin as opposed to the test peptide should not make that much of a difference. It did, particularly since Trp and Met were among them! While it was true that the preformed crosslink was at least in part intact, the total synthesis resulted in an intractable mixture of products. In theory, of course, everything can be separated, but it is important for students to realize that a human life is finite, that a science career is short and that grant renewals are too frequent to battle such odds. The picture of the crude reaction mixture will tell one immediately whether or not a synthesis has been achieved. In this case it was noted rather quickly that at the present stage of protection technology the new process would not work, however attractive it might be in theory. The idea was put on ice pending new developments in the field.

What if one were to produce the prohormone? Of course, a natural prohormone of the size of prorelaxin (160 amino acids)[17] is far beyond the capability of solid phase peptide synthesis which is practically limited to 20 to 40 residues. One hundred residues in one sequence are considered heroic chemistry.[18] Although the complete prohormone cannot be synthesized the principle of single chain synthesis could be utilized to generate relaxin if the pro-piece is shortened. The basis for such a consideration is extracted from the three-dimensional structure of relaxin and insulin which indicates that the N terminus of the A chain and the C terminus of the B chain are in close proximity.[19,20] It had been shown by scientists in the Novo Research Institute in Denmark that insulin would fold properly with a significantly reduced connecting piece (6 amino acids) between the C-terminal end of the B chain and the N-terminal end of the A chain.[21] In fact, insulin will even fold when the N terminus of the A chain is directly connected to the carboxyl group of lysine (B29).[22,23]

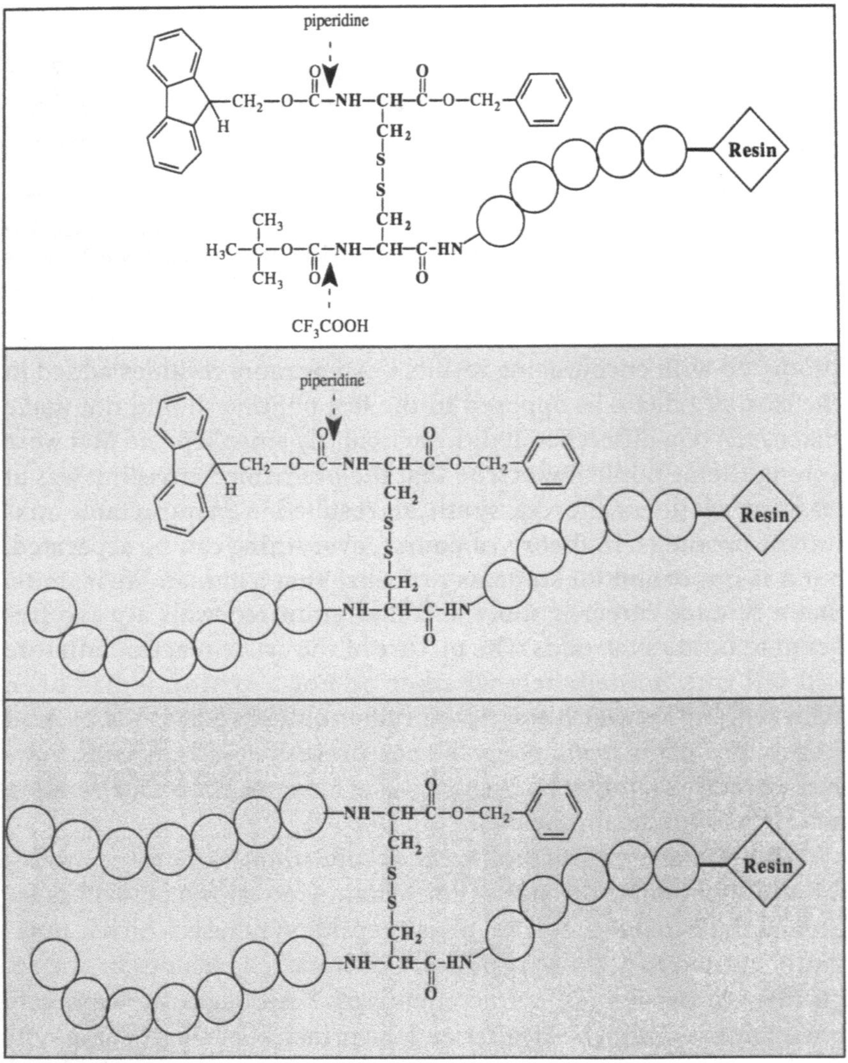

Fig. 8.5. Assembly of an unsymmetrical two-chain molecule on a solid support. Key amino acid of this synthesis is a monovalent unsymmetrical cystine in which one amino group is protected by tert.butyloxycarbonyl (tBoc) and the other by 9-fluorenymethyl-oxycarbonyl (Fmoc). On the Fmoc site the carboxyl group is protected as benzyl ester while the carboxyl group on the tBoc site is free for condensation to a growing polypeptide chain (top panel). Removal of the tBoc group with trifluoroacetic acid allows for the extension of this chain on the tBoc site (middle panel). After the B chain is finished the Fmoc group on the second chain is removed with piperidine and the second chain synthesized with tBoc chemistry (lower panel).

With proper planning relaxin could be generated using a similar concept. The idea was to introduce a six amino acid-connecting peptide which would provide the advantage of prohormone guidance for the disulfide bond formation (Fig. 8.6A,B), and the selective removal of the connecting peptide to produce the active hormone. With such a construct one would have to synthesize a polypeptide of 59 amino acids in length which is not insurmountable and seemed worth a try. The finished single chain prorelaxin was very difficult to dissolve under conditions required for reduction and oxidation. In the end relaxin had not refolded and crosslinked to give a reasonable yield. Further research with this component was discontinued because a more promising synthesis was under development.

Three mild disasters called for a palaver. Our research group huddled for a reassessment meeting, during which liters of coffee were used to flush from our minds and onto paper a new dual approach to the problem. Essentially two groups were formed, one for cloning of a synthetic relaxin-encoding sequence into yeast whereby variants should eventually be produced by site-directed mutagenesis, and the second group was to pursue the chemical synthesis via single chains combined with a yet to be designed sequential and site-selective disulfide bond formation strategy. The groups, of course, were very small such as one or two persons at the most, but after sufficient guessing and teasing as to who would be first to succeed the starting gun went off.

The molecular biology approach made progress fast. The predesigned mini-prorelaxin coding sequence was deduced from the amino acid sequence, small oligonucleotides with overlapping ends were synthesized chemically and ligated enzymatically to produce the coding sequence for human relaxin. Separation and sequence analysis served to find the proper construct so that after three months the gene technologists seemed to have a strong lead.

The most important planning step for a peptide synthesis is to put down the structure with the major functional groups and to develop a strategy. With the previous experience of unconventional approaches it was easy to settle on the procedure involving solid phase synthesis of the A and B chain separately, followed by conversion to the two-chain molecule by appropriate methods. Appropriate, properly defined, evolved into the tactical approach of a synthetic plan. What side chain protections will be used? Which ones will be stable under the conditions prevailing during the synthesis

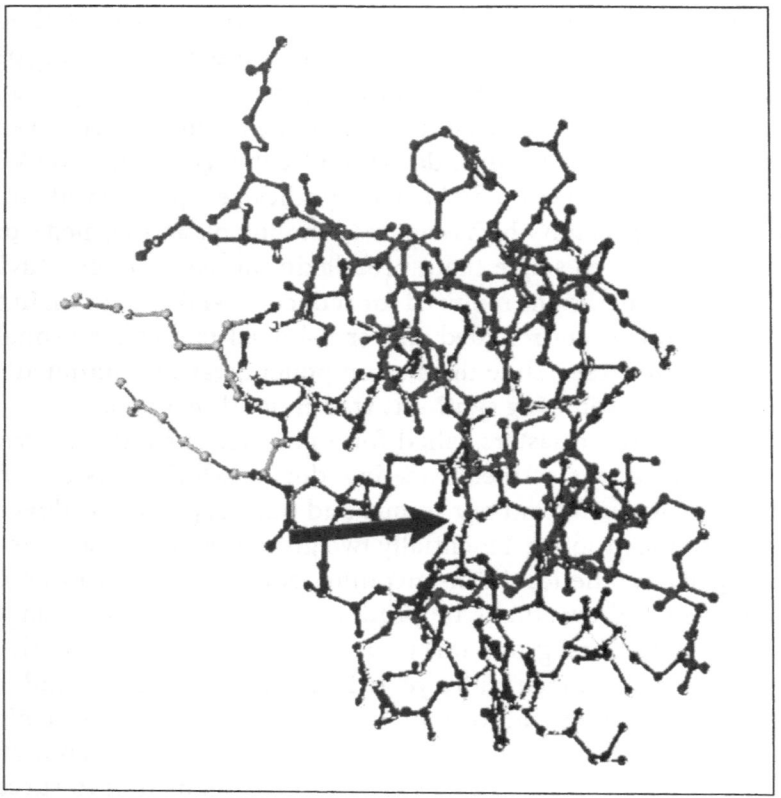

Fig. 8.6. X-ray structure of human relaxin showing the A chain in cyan, the B chain in blue, arginine residues on the B chain helix in white and the artificial connecting peptide in red.

(A, above) Introduction of the peptide Arg-Arg-Glu-Phe-Lys-Arg (red) causes a shift of the N-terminal region of the A chain and the C-terminal region of the B chain but leaves the core structure of the molecule unchanged. The arrow points to the junction of the C terminus of the B chain and the connecting peptide.

(B, opposite) One monomer of the relaxin dimer. (The X-ray data were derived from the Brookhaven databank where they were deposited by Eigenbrot et al 1991). The arrow indicates the point of movement of the C terminus of the B chain in the single chain molecule.

See color insert, pages 191, 192.

at each particular stage, which disulfide bonds should be formed first, and how to prevent other disulfide bonds from being formed inadvertently? The two main methods of synthesizing peptides on a solid phase exist, one involving repeated acid cleavage (tBoc-chemistry) and the other cleavage with a piperidine, an organic base (Fmoc-chemistry).

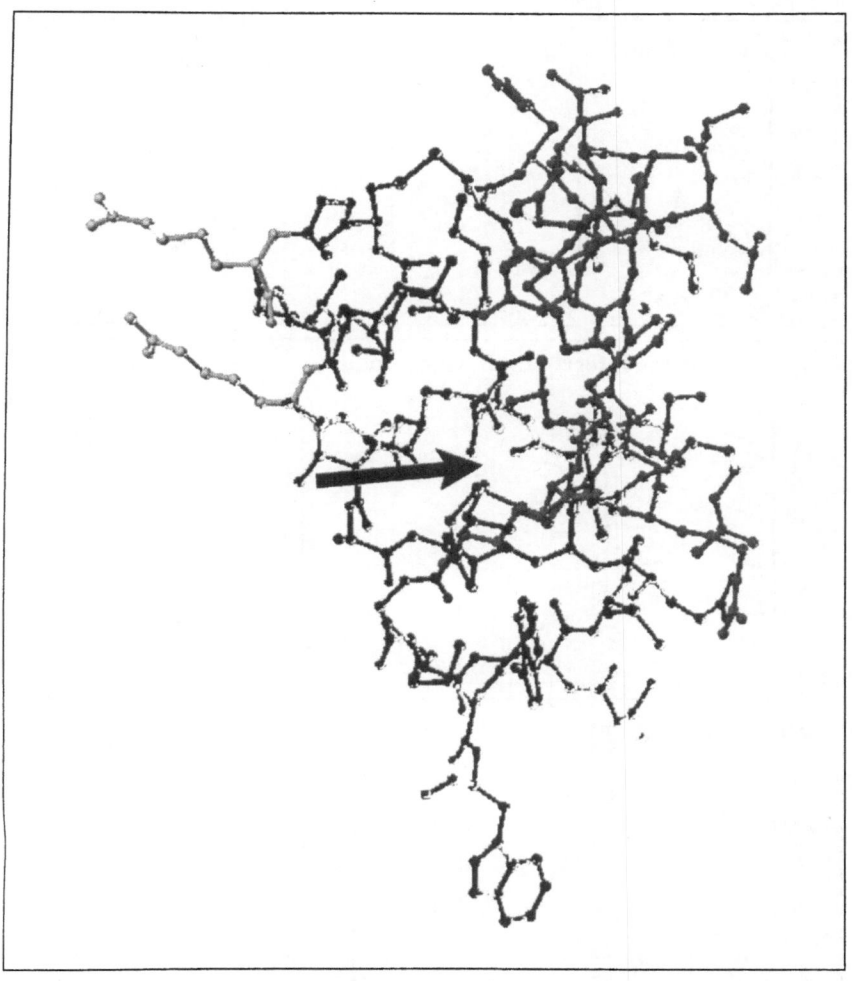

The A chain was synthesized by Fmoc-chemistry with three different sulfhydryl group protections whereby the two SH groups that form the intrachain loop were protected with trityl (Trt) groups. The C-terminal A chain cysteine was protected by the 4-methylbenzyl (MeBzl) group and the remaining interchain crosslink cysteines with the acetamidomethyl (Acm) group. Removal of this peptide from the resin simultaneously caused the trityl groups to release the SH groups which, upon oxidation, formed the intrachain disulfide bond (Fig. 8.7, upper panel). The first interchain disulfide bond to be formed would connect the C-terminal cysteines of both chains and consequently the B chain needed to be finished and purified and ready at this point. The cysteine B11 was protected by Acm, the same group

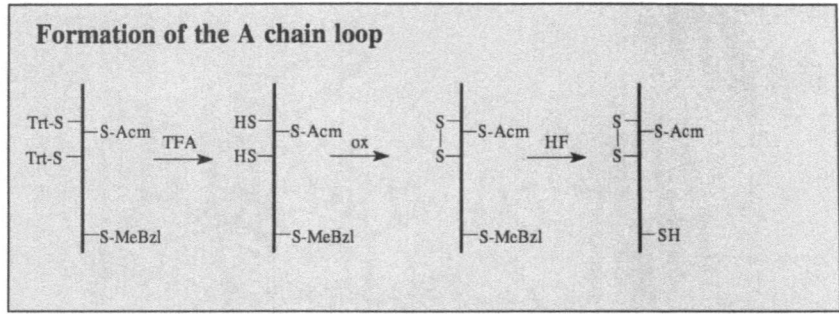

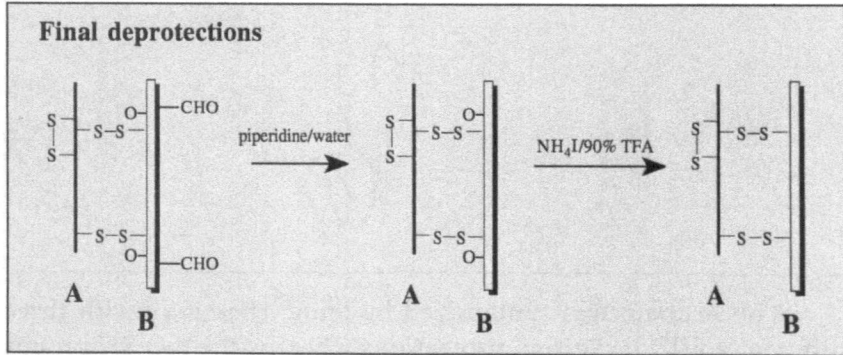

Fig. 8.7. Chemical synthesis of human relaxin.

Upper panel: The relaxin A chain was produced by solid phase peptide synthesis. Only the side chain protection of the cysteines are shown. After treatment with trifluoroacetic acid (TFA) all protecting groups were removed but the acetamidomethyl (Acm) and 4-methylbenzyl (MeBzl) groups. Oxidation of the sulfhydryl groups was achieved by titration with iodine in 50% acetic acid and the A chain purified by preparative HPLC. Liquid hydrogen fluoride in the presence of cresole removed the MeBzl group, releasing a free sulfhydryl group in position A24 which was immediately reacted with the S-activated B chain.

Middle panel: The synthetic B chain was deprotected with HF. Cysteines, methionines and tryptophan side chains remained protected. The S-activating S-[2-(3-nitropyridinesulfenyl)] group reacted with the free sulfhydryl of the A chain in 8 M used to protect the partner SH for crosslink formation in the A chain. The cysteine side chain in B23 was protected as tritylthioether and, after final deprotection with trifluoroacetic acid in the presence of appropriate scavengers, the resulting SH group was activated with 2,2'-dipyridinedisulfide producing a 2-pyridinesulfenyl (PS) group in B23 which will react with a thiol group to form an unsymmetrical disulfide bond even under weakly acidic conditions (Fig. 8.4). The finished B-chain was purified under significant losses. This narration does not include several trials and errors and many of the sacrifices humans make every so often at the altar of stupidity. More than one year had gone by and it was time to visit our molecular biologists.

The problem here was to get the artificial gene or coding sequence into the appropriate vector intact and the right way around. In this discipline the actual experiments are usually finished quickly, but the isolation and recognition of the proper sequence requires a large number of experiments. The coding sequence was incorporated into an expression vector which was used to infect *E. coli*, and positive clones were identified with a radioactive probe which was simply a portion of the artificial gene labeled with ^{32}P. Positive clones were isolated and sequenced and in many trials one or two were found that had the inserted DNA in the proper sequence, i.e., B chain, connecting peptide, followed by A chain so that plans could be made to transfer the coding sequence into a yeast-compatible vector for expression. Yeast could be expected to have the necessary enzymes to activate the prohormone, i.e., to split the artificial connecting peptide once the relaxin had been folded properly and the disulfide bonds

guanidinium chloride at pH 4.5. After 24 h at 37°C the reaction mixture was purified by size-separation followed by preparative HPLC. Thereafter the Acm groups were removed with a 50-fold excess of iodine in 70% acetic acid for 10 minutes at room temperature and the product isolated by preparative HPLC.

Lower panel: Other protecting groups were removed in two steps. Piperidine (10%) in water was used for 2 minutes at room temperature to deprotect the side chains of tryptophan. After acidification and HPLC purification the methionine sulfoxide was reduced with 50- to 100-fold excess of ammoniumiodide in 90% trifluoroacetic acid. The reaction was performed for 15 min on ice and then quenched with aqueous ascorbic acid before separating on an HPLC column.

established. The molecular biologists still felt they were ahead and offered all sorts of consolation and unreasonable bets.

In the camp of peptide chemists the A chain, with its internal disulfide ring complete, was subjected to HF deprotection of the C-terminal cysteine in the presence of the B chain. As soon as the methylbenzyl group was removed the S-(pyridylsulfenyl)-activated SH group of the B chain reacted with the free thiol, forming the second disulfide bond. The product was an A-B crosslinked relaxin with one cysteine in the A chain and one cysteine in the B chain, protected with an Acm group. It took the usual time and effort to extract a small amount of the desired product from all the byproducts. At that point it was clear that conditions had to be changed such as to improve yields. Of course, curiosity drove us to attempt to form the third disulfide bond with iodine in 70% acidic acid (oxidizing conditions). The HPLC chromatogram did not look good and there seemed none of the expected product. As judged by UV spectroscopy the tryptophan side chains were destroyed under those conditions. Although the cysteine-protecting groups were reported to be of limited stability[15] model studies showed that in the presence of free tryptophan oxidation of the indole ring was faster than Acm deprotection. The next synthesis was designed with side chain protection for the indole groups, N(in)-formyl will survive the synthesis via tBoc chemistry and HF deprotection but manipulations with strong nucleophiles, i.e., thiols used for reduction, amines, or high pH will liberate the formyl-protecting group from the tryptophan side chain. The presence of free cysteine or a lysine in the vicinity of tryptophan could cause deprotection of the side chain and destroy all our effort. When these difficulties were finally overcome a way had been opened for the total synthesis of relaxins. The paper of Masueda's group on S-[3-nitropyridinesulfenyl], a new cysteine-protecting and activating group,[24] showed exactly what was needed, i.e., an S-activated cysteine compatible with tBoc chemistry. By incorporating this cysteine residue in position B23 the chain could be synthesized with tryptophan and its protecting group intact. The B chain was soluble in saturated guanidinium chloride at pH 4.5 but all other solvents failed which left us with size-separation as the only purification procedure. The synthesis of the first interchain disulfide bond went as well as before whereas the formation of the last disulfide in the presence of an excess of iodine still gave a variety of peaks containing some of the fully disulfide-linked relaxin with N(in)-formyl protecting groups on the tryptophans. Other fractions, however, showed in addition oxidized thioether groups in methionine. Treatment with 10% piperidine in water removed the formyl groups from the tryptophan side-chains. At the end there was a little impure product left, and while it did not satisfy our peptide group at all, it was enough to tease the molecular biologists who could only show that the proper coding sequence had been incorporated into the yeast plasmid, but no product could be detected in the growth medium.

The relaxin coding sequence had been incorporated behind the alpha-pheromone promoter and relaxin should therefore have been secreted into the medium. Several other promoters were tried for the next year without leading to more than just enough material for a bioassay and radioimmunoassays.[25]

Since the removal of the Acm groups caused a partial oxidation of methionine the big question was whether or not one can reduce methionine sulfoxide in the presence of disulfide bonds. According to the literature one could,[26] but it seemed impossible to find the most relevant article in the library. Impatiently the reaction was tried according to general principles. It worked! When the article arrived from the interlibrary loan desk we read to our amusement that the reaction does not work when tryptophan is involved.[27] Success in this case was probably due to the difference in the primary and secondary structure of the peptide or simply the chemical workup of the product mixture. Students should not mistake this for an invitation to ignore the literature; analogies are just not perfect when it comes to peptides. During the reaction methionine sulfoxide is reduced to methionine and iodide is oxidized to iodine (Fig. 8.8). Reducing the generated iodine with ascorbic acid followed by HPLC separation did fine for us, while Izeboud and Beyerman[27] observed multiple products when they reduced the iodine with thiosulfate. Subsequently relaxins were synthesized with methionine sulfoxide, and the sulfoxide reduced in the last step with ascorbate, and that established the first good synthesis (Fig. 8.7).[28]

It was two to three years after the starting gun went off and by this time the chemistry group had polished up the synthesis by slight changes in condition and reaction times and by developing the touch for handling the chains, a quality that can never be described in the literature, and had increased the yield to between 2% (which is very significant) and 25% (which is outstanding) depending upon which molecule had been synthesized. Low yields were usually affiliated

$$R-\underset{\underset{O}{\|}}{S}-CH_3 \ + \ I^- \ + \ 2H^+ \ \rightleftharpoons \ R-S-CH_3 \ + \ H_2O \ + \ 1/2\,I_2$$

Fig. 8.8. Reversible reduction of methionine sulfoxide under acidic conditions with iodide.

with poor solubility of the B chain, limiting the final purification of this chain. Whenever HPLC-purified B chain and A chain were used, yields were in the range of 15 to 25%. This is true for insulin and insulin analogs, bombyxin, and relaxins from mouse, Syrian hamster, and guinea pig. Within a month or two one could make any desired derivative and could substitute amino acids in the center of the chain as well as at the periphery. Up to that time most of the structure/function work for insulin, for example, had been limited to semisynthesis either by chain combination of a synthetic modified chain with a chain derived from native insulin, the exchange of the C-terminal octapeptide in the B chain,[29-31] or additions or deletions at the N-terminal end of the A or B chain.[32,33] Some substitutions at the C-terminal end of the A chain had been made with various cumbersome tricks,[34] but none of this freewheeling direct production of any desired derivative had been practiced before. Rittel and his colleagues[35] at Ciba Geigy had actually set up a similar insulin synthesis which never became popular because all chemical reactions were performed in solution and the synthesis was labor-intensive and time-consuming. Nevertheless, without their pioneering work the total synthesis of relaxin would have been even more difficult. We believe that no scheme in molecular biology can come close to the speed and versatility with which these small proteins can now be produced by solid-phase peptide chemistry. In contrast, the majority of insulin analogs have been made by site-directed mutagenesis from bioengineered insulin precursors (proinsulin or mini-insulin) either in *E. coli* or in yeast.[36-38]

The story is quite different if one wants to produce a large amount of one kind, i.e., grams instead of milligrams, or if one deals with very large proteins that are outside the range of peptide chemistry, then molecular biology is certainly the only method of choice. Two big advantages of chemistry are that derivatives which are really of interest to protein chemistry can be forced into the proper configuration by selective sequential disulfide bond formation and that non-native amino acids such as citrulline can be incorporated with ease. If the molecular biologists produce variants through site-directed mutagenesis in such a small molecule the folding is often severely disturbed and there are several cases in the literature where modifications have led to denatured products that could not be excreted by the cell.

During the first three quarters of this century peptide chemistry was a very exciting field with pioneering efforts such as the first semi- and total syntheses of proteins.[39-42] When it comes to large proteins, the technology of molecular biology is justifiably replacing chemistry. As a consequence the skills of peptide chemistry are no longer known enough to come to mind at once in cases where that technique would be clearly superior. As human affairs proceed, this too will change again and already some work on efficient methods of fragment condensation have been reported which may move proteins back into the purview of peptide chemists.[43]

References

1. Walder JA, Walder RY, Heller MJ et al. Complementary carrier peptide synthesis: General strategy and implications for prebiotic origin of peptide synthesis. Proc Natl Acad Sci USA 1979; 76:51-55.
2. Ten Kortenaar PBW, Van Dijk BG, Peeters JM et al. Rapid and efficient method for the preparation of Fmoc-amino acids starting from 9-fluorenylmethanol. Int J Peptide Protein Res 1986; 27:398-400.
3. Merrifield RB. Solid phase peptide synthesis: I The synthesis of a tetrapeptide. J Am Chem Soc 1963; 85:2149-2154.
4. Atherton E, Sheppard RC. Solid phase peptide synthesis, a practical approach. Oxford: IRL Press, 1989. (Rickwood D, Hames BD, eds. Practical Approach Series;
5. Stewart JM, Young JD. Solid phase peptide synthesis. Rockford, IL: Pierce Chemical, 1984.
6. Katsoyannis PG, Tometsko A. Insulin synthesis by recombination of A and B chains: A highly efficient method. Proc Natl Acad Sci USA 1966; 55:1554-1561.
7. Gattner HG, Krail G, Danho W et al. Eine verbesserte Methode der Kombination von Insulinketten zur Darstellung von Insulinanalogen. Hoppe Seyler's Z Physiol Chem 1981; 362:1043-1049.
8. Canova-Davis E, Baldonado IP, Teshima GM. Characterization of chemically synthesized human relaxin by high performance liquid chromatography. J Chromatogr 1990; 508:81-96.
9. Steiner DF, Clark JL. The spontaneous oxidation of reduced beef and rat proinsulin. Proc Natl Acad Sci USA 1968; 60:622-629.
10. Steiner DF, Rouille Y, Gong Q et al. The role of prohormone convertases in insulin biosynthesis: Evidence for inherited defects in their action in man and experimental animals. Diabetes & Metabolism 1996; 22:94-104.

11. Halban PA. Proinsulin processing in the regulated and the constitutive secretory pathway. Diabetologia 1994; 37:S65-72.
12. Wittinghofer A. Synthese der Schafinsulin-A-Kette mit 6-11-Disulfidring. Liebigs Ann Chem 1974; :290-305.
13. Sieber P, Kamber B, Eisler K et al. 158 Synthese von Humaninsulin II Aufbau des cyclischen Fragments A(1-13). Helv Chim Acta 1976; 59:1489-1497.
14. Birr C, Pipkorn R. Voll aktives Insulin durch selektive Bildung der Disulfidbrücken zwischen synthetischer A-Kette und natürlicher B-Kette. Angew Chem 1979; 91:571-573.
15. Kamber B, Hartmann A, Eisler K et al. 96 The synthesis of cystine peptides by iodine oxidation of S-trityl-cysteine and S-acetamido-methyl-cysteine peptides. Helv Chim Acta 1980; 63:899-915.
16. Büllesbach EE, Schwabe C. Sequential synthesis of an unsymmetrical two-chain disulfide peptide on solid support. Tetrahedron Lett 1992; 33:5881-5884.
17. Kemp BE, Niall HD. Relaxin. Vitamins and Hormones 1984; 41:79-115.
18. Kent SBH. Chemical synthesis of peptides and proteins. Ann Rev Biochem 1988; 57:957-989.
19. Baker EN, Blundell TL, Cutfield JF et al. The structure of 2Zn pig insulin crystals at 1.5A resolution. Phil Trans R Soc Lond B 1988; 319:369-456.
20. Eigenbrot C, Randal M, Quan C et al. X-ray structure of human relaxin at 1.5 Å: Comparison to insulin and implications for receptor binding determinants. J Mol Biol 1991; 221:15-21.
21. Thim L, Hansen MT, Norris K et al. Secretion and processing of insulin precursors in yeast. Proc Natl Acad Sci USA 1986; 83: 6766-6770.
22. Markussen J, Jørgensen KH, Sørensen AR et al. Single chain des-(B30)insulin: Intramolecular crosslinking of insulin by trypsin catalyzed transpeptidation. Int J Peptide Protein Res 1985; 26:70-77.
23. Derewenda U, Derewenda Z, Dodson EJ et al. X-ray analysis of a single chain B29-A1 peptide-linked insulin molecule: A completely inactive analogue. J Mol Biol 1991; 220:425-433.
24. Bernatowicz MS, Matsueda R, Matsueda GR. Preparation of Boc-[S-(3-nitro-2-pyridinesulfenyl)]-cysteine and its use for unsymmetrical disulfide bond formation. Int J Peptide Protein Res 1986; 28:107-112.
25. Yang S, Heyn H, Zhang YZ et al. The expression of human relaxin in yeast. Arch Biochem Biophys 1993; 300:734-737.
26. Beyerman HC, Izeboud E, Kranenburg P et al. Synthesis of methionine-containing peptides via their sulfoxides. In: Gross E, Meienhofer J, eds. Sixth American Peptide Symposium. Pierce Chemical Co. Rockford Il, 1979:333-336.
27. Izeboud E, Beyerman HC. Synthesis of substance P via its sulfoxide by the repetitive excess mixed anhydride (REMA) method. Recl Trav Chim Pays-Bas Neth 1978; 97:1-6.

28. Büllesbach EE, Schwabe C. Total synthesis of human relaxin and human relaxin derivatives by solid phase peptide synthesis and site-directed chain combination. J Biol Chem 1991; 266:10754-10761.

29. Nakagawa SH, Tager HS. Role of the phenylalanine B25 side chain in directing insulin interaction with its receptor: Steric and conformational effects. J Biol Chem 1986; 261:7332-7341.

30. Casaretto M, Spoden M, Diaconescu C et al. Shortened insulin with enhanced in vitro potency. Biol Chem Hoppe Seyler 1987; 368: 709-716.

31. Mirmira R, Nakagawa SH, Tager HS. Importance of the character and configuration of residues B24, B25, and B26 in insulin-receptor interactions. J Biol Chem 1991; 266:1428-1436.

32. Geiger R, Geisen K, Summ HD et al. [A1-D-alanine]insulin. Hoppe Seyler's Z Physiol Chem 1975; 356:1635-1649.

33. Nakagawa SH, Tager HS. Importance of aliphatic side-chain structure at position 2 and 3 of the insulin A chain in insulin-receptor interactions. Biochemistry 1992; 31:3204-3214.

34. Gattner HG, Schmitt EW. [A21-Asparaginimide] insulin. Saponification of insulin hexamethyl ester, I. Hoppe-Seylers Z Physiol Chem 1977; 358:105-113.

35. Sieber P, Kamber B, Hartmann A et al. Totalsynthese von Human insulin: Beschreibung der Endstufen. Helv Chim Acta 1977; 60:27-37.

36. Brange J, Owens DR, Kang S et al. Monomeric insulins and their experimental and clinical applications. Diabetes Care 1990; 13: 923-954.

37. Brems DN, Brown PL, Bryant C et al. Improved insulin stability through amino acid substitution. Protein Eng 1992; 5:519-525.

38. Hoogwerf BJ, Mehta A, Reddy S. Advances in the treatment of diabetes mellitus in the elderly. Development of insulin analogues. Drugs & Aging 1996; 9:438-448.

39. Wieland T, Bodanszky M. The World of Peptides: A Brief History of Peptide Chemistry. Berlin: Springer Verlag, 1991.

40. Hofmann K, Smithers MJ, Finn FM. Studies on polypeptides. XXXV. Synthesis of S-peptide 1-20 and its ability to activate S-protein. J Am Chem Soc 1966; 88:4017-4019.

41. Hirschmann R, Nutt RF, Veber DF et al. Studies on the total synthesis of an enzyme. V. The preparation of enzymatically active material. J Am Chem Soc 1969; 91:507-508.

42. Gutte B, Merrifield RB. The synthesis of ribonuclease A. J Biol Chem 1971; 246:1922-1941.

43. Dawson PE, Muir TW, Clark-Lewis I et al. Synthesis of proteins by native chemical ligation. Science 1994; 266:766-779.

CHAPTER 9

Analytic of the Prototype

It is time to interrupt the main scenario now for an extensive description of the procedures required to demonstrate the purity of synthetic human relaxin. The physiologists should particularly appreciate a rigorous approach to this portion of the research for if there are questions about the purity or authenticity of a hormone, enzyme, factor or whatever, unequivocal interpretation of their measurements in whole animals would be even more difficult. Relaxin obliges to provide an example of such an occurrence which casts a shadow over many years of painstaking measurements in whole animals. This chapter is set aside for important technical know-how that is required to ascertain authenticity and purity of the product of a synthesis. Students of this discipline might think of it as a protein chemists' bootcamp, designed to make him exert a great deal of effort to destroy his illusions about a "pure" product.

The first concern is to verify homogeneity of the polypeptide. Complexity of these molecules and the limits of each separation system require more than a single procedure to prove the identity of the target structure.[1] High performance liquid chromatography on reversed phase columns[2,3] is a high power separation tool for purification and isolation, yet proving homogeneity in a chromatography system requires high-technology analytical systems and solvent systems and columns different from those used for the isolation procedure. Electrophoresis in different buffers will provide an additional proof of purity, particularly if none of the prior purification steps involved separation on ion exchange columns.

Once the relaxin appears as a single peak in electrophoresis and analytical reversed phase HPLC (Fig. 9.1), component chains need to be identified. Reduction, followed by immediate transfer of the

Relaxin and the Fine Structure of Proteins, by Christian Schwabe and Erika E. Büllesbach. © 1998 Springer-Verlag and R.G. Landes Company.

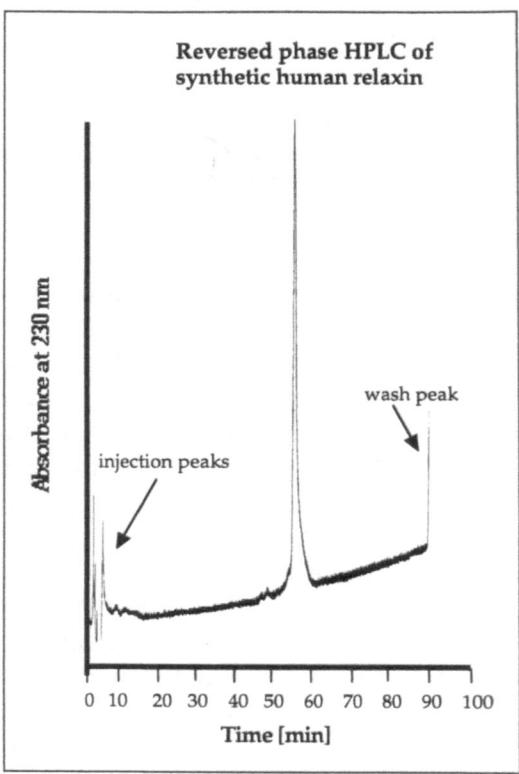

Fig. 9.1. Analytical HPLC on an Aquapore 300 (C_8, 2.1 mm x 30 mm) column. The solvent system consisted of 0.1% trifluoroacetic acid in water (A) and 0.1% trifluoroacetic acid in 80% acetonitrile. The protein (1 µg of relaxin) was injected and eluted with a linear gradient from 25% B to 45% B in 90 minutes at a flow rate of 100 µl/min. The column was washed with 100% B for 5 minutes before equilibrating with 25% in preparation for the next run.

mixture to an HPLC column, should show component polypeptides with retention times different from each other and from the intact molecule (Fig. 9.2).

Acid hydrolysis of the intact molecule and the isolated chains, followed by a quantitative amino acid analyses, implies that chains were present in the expected ratio in the intact molecule (Table 9.1). Amino acid and sequence analysis serve to identify the component amino acids and assure that they are in the right place in each of the chains. Certain amino acids are destroyed in part by acid hydrolysis, i.e., tryptophan and cysteine, whereas glutamine, asparagine and pyroglutamic acid are hydrolyzed to the corresponding dicarboxylic acids. Keeping these restrictions in mind, amino acid analysis will provide important information concerning the relative amounts of amino acids in each chain. Sequence analysis by automatic Edmann degradation assures one that each amino acid is in the right position relative to all other amino acids. Since the synthesis of a proto-

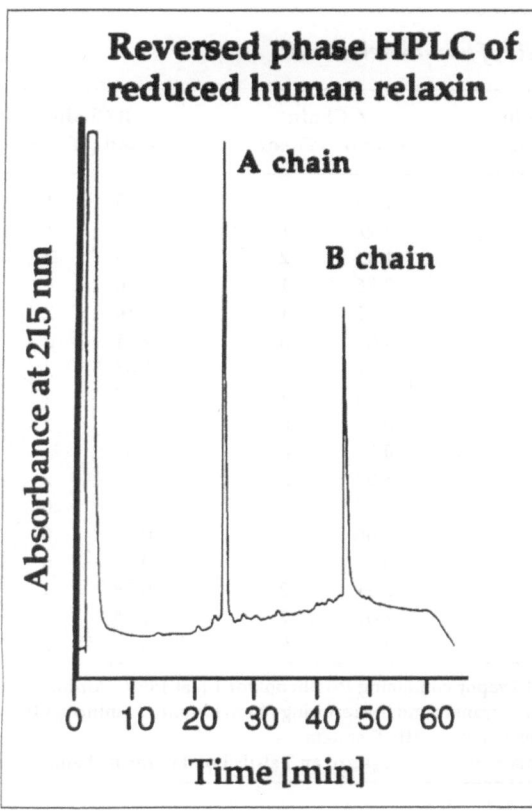

Reversed phase HPLC of reduced human relaxin

A chain

B chain

Absorbance at 215 nm

0 10 20 30 40 50 60

Time [min]

Fig. 9.2. Human relaxin (2 µg) in 50 µl water was reduced with 50 µl of 50 mM dithiothreitol in 0.2 M Tris/HCl pH 8.6 in 6 M guanidinium chloride for 1 h at 37°C. The reaction was acidified and separated on Aquapore 300 applying a linear gradient from 10% to 65% B in 60 minutes at a flow rate of 100 µl/min. (Chromatography system see Fig. 9.1).

type is a crucial initial step for further structure/function- or physiological investigation, sequence analysis is necessary to exclude possible errors in the assembly of the molecule.

Enzymatic digest of the unreduced relaxin, followed by HPLC peptide mapping and subsequent hydrolysis and amino acid analysis of the fractions will reveal the proper location of the disulfide bonds and the parallel A and B chain orientation (Fig. 9.3). After the first synthesis a tryptic digest was followed by synthesis of all possible fragments by chemical routes different from those used for the total synthesis. All fragments were found again in the tryptic digest of human relaxin II, indicating that no chemical modifications had occurred. Due to low UV absorbance some of the smaller fragments (free arginine, and the two tetra-peptides) cannot be identified easily and may disappear in the background noise. A combination of different chromatographic procedures and chemical fragmentation has to fill the gaps left by each method.

Table 9.1. Amino acid analysis of human relaxin II

	Human Relaxin		A Chain		B Chain	
	Found	Theor.	Found	Theor.	Found	Theor.
Asp	2.04	2	1.05	1	0.90	1
Thr	1.93	2	0.97	1	1.03	1
Ser	4.56	5	1.85	2	2.95	3
Glu	4.67	5	0.96	1	3.76	4
Gly	3.01	3	0.99	1	2.19	2
Ala	4.80	5	3.00	3	2.11	2
Cys*	3.72	6	2.51	4	1.43	2
Val^	2.40	3	1.03	1	1.44	2
Met	1.86	2	0	0	2.01	2
Ile^	2.30	3	0	0	2.39	3
Leu	5.20	5	3.07	3	2.00	2
Tyr	1.10	1	0.82	1	0	0
Phe	0.98	1	1.06	1	0	0
His	0.90	1	0.87	1	0	0
Lys	3.03	3	2.01	2	0.78	1
Arg	4.00	4	2.01	2	1.65	2
Trp*	0	2	0	0	0	2

Hydrolysis was performed in 6 N HCl vapor containing 1% phenol for 1 h at 150°C. Amino groups were tagged with phenylisothiocyanate and the resulting phenylthiourea-amino acids separated on a Waters' Pico Tag reversed phase HPLC system.
*destruction during hydrolysis; ^ incomplete hydrolysis of an Val-Ile bond in the B chain.

The integrity of tryptophan must be investigated by UV spectroscopy (Fig. 9.4). The typical UV spectrum of a tyrosine-, tryptophan-, and phenylalanine-containing peptide can be used for the quantitative determination of the protein content. Since a quantitation of amino acids is also possible by amino acid composition, the two methods are complementary and the results should match. In addition the UV spectrum of the protected N(in)formyl tryptophan is quite different from unprotected tryptophan and completeness of the deprotection can be verified.[4] Oxidative destruction of tryptophan only rarely yields a single product but their presence can be excluded by UV spectroscopy.

Oxidation of methionine is not readily detected by any of these methods because the sulfoxide is converted to methionine upon acid hydrolysis as well as during sequence analysis. The relative value ordinarily observed during quantitative amino acid hydrolysis tends to be low for methionine. Sulfoxide formation can be detected by

HPLC analysis because increased hydrophilicity accelerates elution. Complete sulfoxide formation can be detected by peptide mapping after tryptic digestion, followed by comparison of the fragments with independently synthesized peptides. The best way to detect sulfoxides is by mass spectrometry (Fig. 9.5)[5] which can now be done with a relatively inexpensive "time-of-flight" (TOF) instrument.

Once a prototype has been thoroughly characterized, analogs require less stringent analyses. Parallel analytical experiments of the new analog and the initial molecule will give a good indication about the identity of the analog. Retention times in reversed phase HPLC of the intact analog will most likely differ from that of the initial molecule but after reduction only the chain carrying the modification will be different from its partner chain. Enzymatic digest and peptide mapping of the intact derivative will indicate only one peptide that differs between the two molecules. This peptide could be sequenced to verify the correctness of the analog. There should be no other deviation but if there is one, it means a lot of analytical work and troubleshooting to find out what went wrong.

If this seems like a lot of work; it is! To boot, the editors put it into small print if it is not deleted outright, and in a grant application it goes into the unread section called appendix. The obvious question is why do it? Technically the analytic tells that one really has done what one set out to do, and that has always motivated humans. In the case of biologically important molecules, however, the slightest doubt about the structural correctness of the molecule is unacceptable if the bioassay results are to be interpretable.

Building from chemicals on the shelf something that interacts predictably and beneficially with a system that has been produced spontaneously over billions of years gives such efforts an almost poetic quality. This book, as the reader must have noticed, celebrates that most enjoyable irrationality. The pleasures one may experience as a consequence of this "chat with nature", may be more readily forgiven by society if the product also provides the pleasures of improved lives for others.

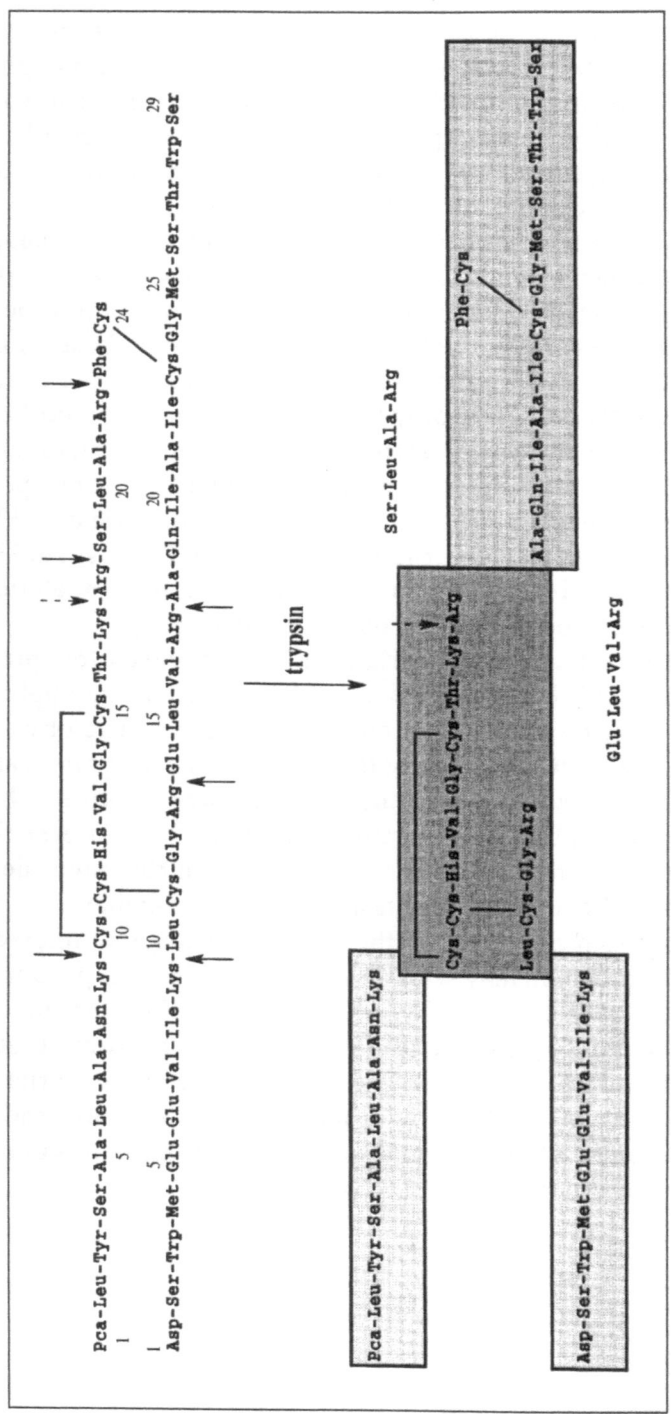

Fig. 9.3A. Upper panel shows the primary structure of human relaxin II with arrows indicating the potential tryptic cleaving sites. The four large tryptic fragments are indicated. Relaxin was digested with trypsin at an E:S ratio of 1:25 in 50 mM ammonium hydrogen carbonate for 1 h at 37°C.

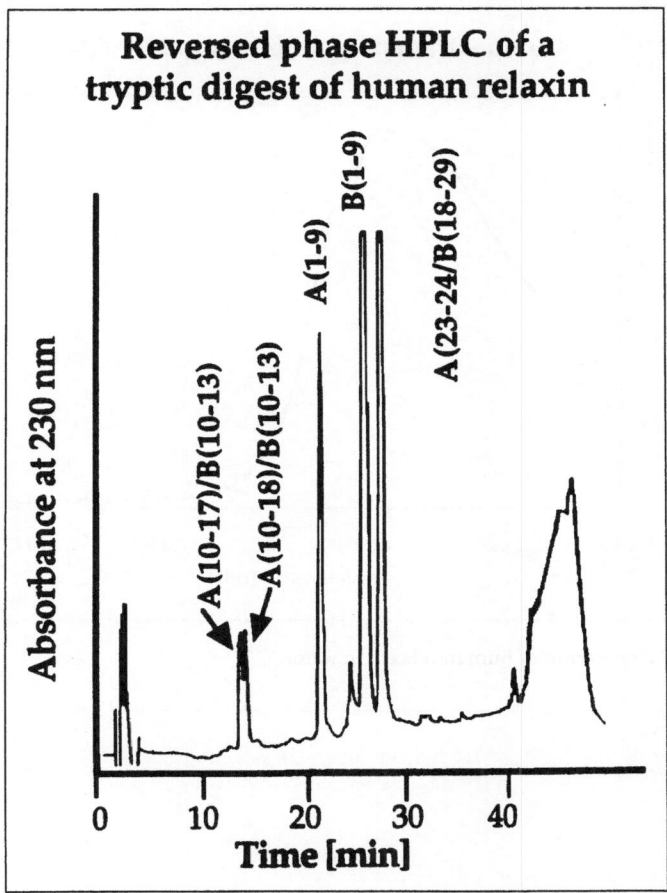

Fig. 9.3B. The reaction mixture was acidified with 0.1% trifluoroacetic acid and separated on Aquapore 300 using a linear gradient from 0 to 45% B in 40 min at a flow rate of 100 µl/min (The chromatography system used is described in Fig. 9.1).

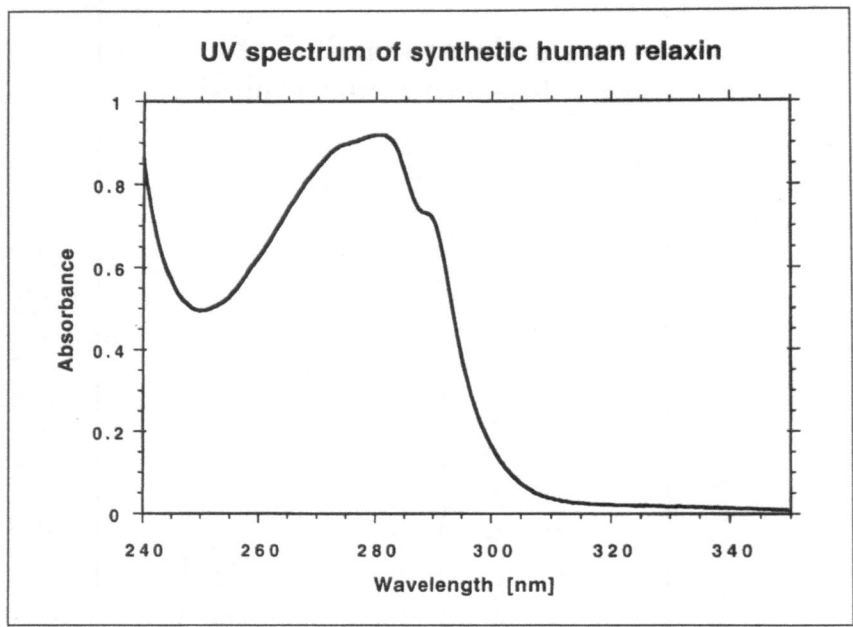

Fig. 9.4. UV spectrum of human relaxin in water.

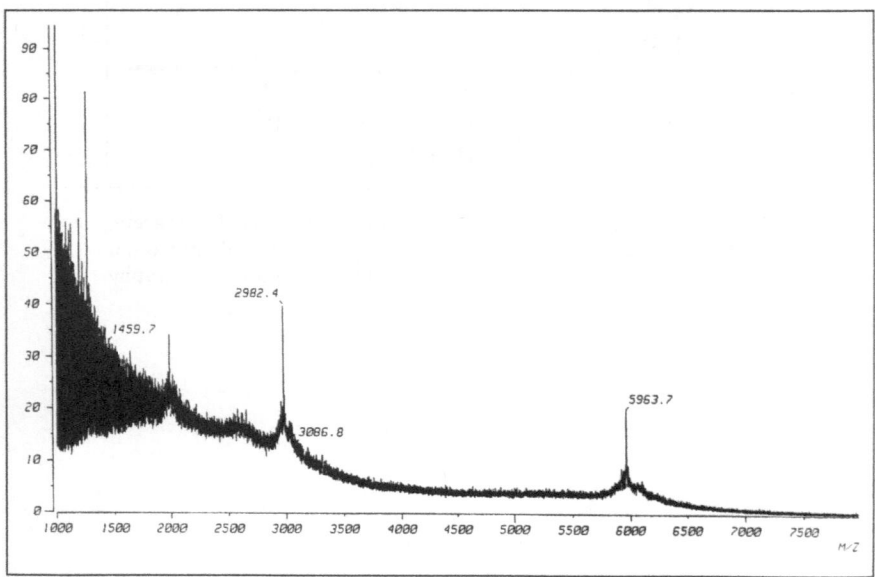

Fig. 9.5. Mass spectrometry of human relaxin on a JOEL 4-tandem mass spectrometer.

References

1. Karger BL, Hancock WS, eds. High Resolution Separation and Analysis of Biological Macromolecules. San Diego: Academic Press, 1996. (Abelson JN, Simon MI, ed. Meth Enzymology; vol 270).
2. Aguilar MI, Hearn MTW. High resolution reversed-phase high performance liquid chromatography of peptides and proteins. Meth Enzymology 1996; 270:3-26.
3. Mant CT, Hodges RS. Analysis of peptides by high-performance liquid chromatography. Meth Enzymol 1996; 271:3-50.
4. Büllesbach EE, Schwabe C. Total synthesis of human relaxin and human relaxin derivatives by solid phase peptide synthesis and site-directed chain combination. J Biol Chem 1991; 266:10754-10761.
5. Yates JR. Protein structure analysis by mass spectrometry. Meth Enzymol 1996; 271:351-377.

CHAPTER 10

The Receptor-Binding Site of Relaxin

The new chemical synthesis of human relaxin finally provided the tool to deny or confirm earlier observations with CHD-modified relaxin, and if the basic idea was correct one should be able to obtain unequivocal answers as to which of the arginines is important if not both, or whether merely a positive charge is required. Another important question pertains to the distance between the core of relaxin and the guanidino group which could be answered by incorporation of homoarginine at the binding-site.

The contenders for the first substitution are shown in Figure 10.1. The unnatural amino acid citrulline is isosteric with arginine except that the substitution of a urea for a guanidino group eliminates the positive charge.

The first human relaxin analog synthesis involved the displacement of arginine B13 and B17.[1] Nothing unusual happened during the synthesis except that we had to replace Boc-Arg(Tos) (Tos is the abbreviation for tosyl, the arginine side chain-protecting group) by Boc-citrulline in the appropriate synthesizer cartridges. Since every residue presents a different problem for coupling, the synthesis was stopped after each addition of the unusual amino acids and checked for completeness. This test called Kaiser test, according to its originator, requires that a small resin sample be removed from the reaction vessel, dried, weighed and then reacted with ninhydrin to determine the unreacted amino groups by measuring spectrophotometrically the blue color intensity.[2,3] The darker the mixture the more unreacted amino groups and, consequently, the poorer the coupling yield. Bad yields are corrected by double coupling. The citrulline B chains were purified and combined with normal synthetic A chains precisely as described in the previous chapter.

Relaxin and the Fine Structure of Proteins, by Christian Schwabe and Erika E. Büllesbach. © 1998 Springer-Verlag and R.G. Landes Company.

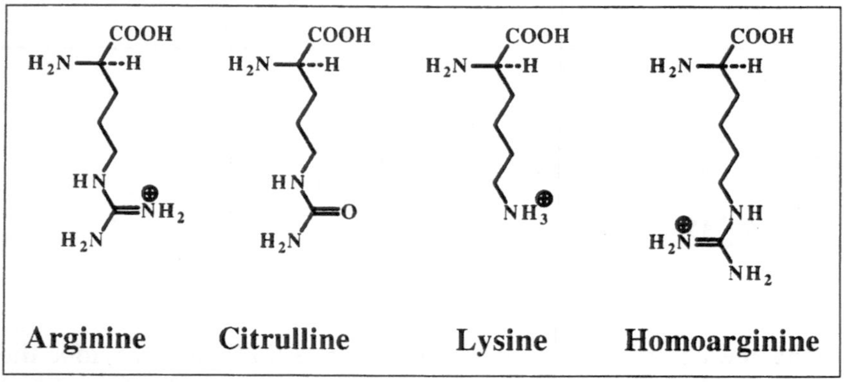

Fig. 10.1. Comparison of the structures of arginine and other amino acids used to replace arginine in the active site of relaxin.

Finally the day arrived for the *experimentum crurum*. As described earlier, ovariectomized young female mice had been injected with a depot form of estrogen and after five days they received injections of 0.4, 0.8 or 1.2 micrograms of relaxin, 1.0, 5.0 or 20 micrograms of citrulline relaxin. A control group received only the carrier, benzopurpurin-4B (Fig. 10.2). The animals were left overnight and the measurements started as the first thing next morning. Only one person of our group knew which animals had received what preparation and the dissectors as well as the one who made and recorded the actual measurements under the dissecting scope were left in the dark.

The tension and excitement was hanging in the air but nobody said very much. Occasional mutterings from the measurer interrupted the silence with words like fully active, active again, nothing here, very active, active again, where is our citrulline? Active groups were interspersed with one or two inactive ones but nobody knew at this point whether these were low concentrations of standard or whether our modification of the arginines had indeed eliminated bioactivity. The one who knew had a stone face until the end of the measurements. When the code was broken we had discovered the receptor-binding site of relaxin!

There was a modest celebration. After all nobody could pinpoint a receptor binding-site for insulin with the same kind of accuracy. The sobering idea that this may be an inherent property of hormones some of which may have somewhat diffuse, others very

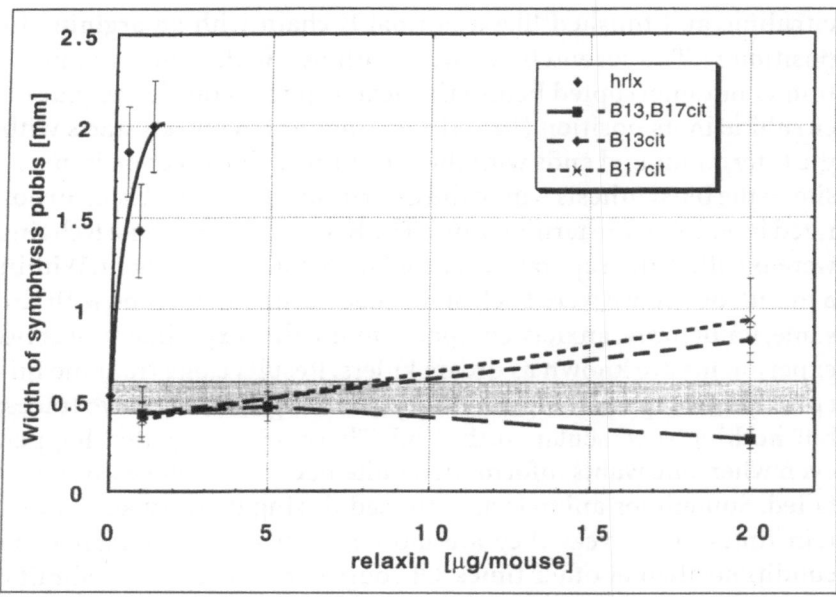

Fig. 10.2. Dose response of human relaxin (hrlx), bis-citrulline(B13,B17) human relaxin (B13,B17Cit), citrulline(B13) human relaxin (B13Cit) and citrulline(B17) human relaxin (B17Cit). The gray area indicates the width of the symphysis pubis of the negative control group. The bars represent standard errors.

sharply defined binding sites, did not crimp our feeling that we had surpassed the insulin people who, after all, had a 50-year head start and were hundreds of investigators strong.

The citrulline relaxin was inactive at any significant concentration as compared to the control within the same experiment (Fig. 10.2). Returning one by one from a delirious state it was decided to find which of the two arginines B13 or B17 would be important or whether both are required. All eyes wandered over to our peptide chemists. It is still a hell of a job, let the molecular biologists do it! That was a snide remark in light of the fact that the molecular biologist had been unable to overcome the poor expression yield of relaxin and had been left alone with a problem for which industry usually employs groups of 10 or more investigators. Expression of products in the proper form, we learned by experience, is the more difficult of all the problems of molecular biology.

Two B chains had to be synthesized and to expedite the work a synthesis was interrupted just before addition of the first arginine (B17). At this point the resin was divided, one-half was reacted with citrulline and finished like a normal B chain with an arginine in position 13. The second batch was continued with arginine in position 17 but interrupted before the next arginine addition to place a citrulline in its position (B13) (Note: chemical synthesis starts with the C terminus and ends with the N terminus, the direction is opposite to the biosynthesis where the C-terminal amino acid is incorporated last just before termination). The B13 and B17 citrulline B chains were purified and separately reacted with a normal A chain. Within a month or two we were back at the measuring microscope with the same, if not more, anxiety compared to the first experiment. Second experiments are known as dream killers. Results came from the microscope: Group 1 nothing; 2 nothing; 3 nothing; 4 a trace; 5 two traces; but nothing spectacular to the end. These are things that happen even when one wants information quite badly. The whole assay had failed. Sometimes animals are stressed during delivery and at certain times of the year they are more sensitive to such changes in conditions than at other times. Of course one would never admit a weighing error when biology could be blamed so easily. Concentrations in the microgram range were really never made up by weight but rather by UV absorbance of stock solutions.

The assay was tried again, fourteen days lost. Group one nothing, group two good activity, group three nothing, group four full activity; great, at least this assay worked. Group 5, 6, 7, 8, 9 and 10, no activity. When our peptide chemist looked at the results she said, oh, that looks good and this was the utmost exuberance that could ever come from that member of our group. There was no doubt, both arginines were required. B13 citrulline was as inactive as B17 citrulline and only slightly better at high doses than B13, B17 bis-citrulline (Fig. 10.2). Competitive binding to the relaxin-receptor in vitro required a concentration that was three orders of magnitude higher than human relaxin. The interaction with the receptor (Fig. 10.3) could be visualized like the connection of an electrical plug and receptacle.[4] This was a very unusual mode of hormone receptor interaction and there was no equivalent in the literature. How important this finding really was we began to appreciate when we failed to get it into print in a couple of interdisciplinary journals. It seemed appropriate to publish where protein structure researchers would look as well because this was not relaxin-specific work alone but it also carried a message for the general principle of receptor hormone interaction that might be of interest to groups working on a completely different aspect of the endocrine system and who might, like we at first, shy away from implicating arginine as a binding-site residue. Most investigators are familiar with the finer nuances of journal-investigator relations, but if students are nearly as starry-eyed as these authors were at the beginning of their careers they may have a hard time understanding the games that grown ups play. It is an age-old problem however. Szent György once remarked that if nobody likes your paper at first you know you have something significant, and Rosalyn Yallow undoubtedly inspired by her own experience made the classical statement that the significance of your paper is inversely proportional to the ease with which you can publish.

We had been immodest in our paper, suggesting that this was indeed the first example of this dual prong mechanism of binding involving arginine residues (which it was). The reviews were mixed but the editor chose to pay attention to the one that cited irrelevant examples where some basic amino acid was closed to a receptor-binding site. The paper was eventually published in a less visible spot[4] but still those who should eventually become important for our progress and success found it and took notice.

The guanidino group is one of the strongest organic bases in biology and it was natural to wonder whether that group was indeed required or whether a simple positive charge like the ε-amino group of lysine would be sufficient. A B13, B17 lysine relaxin was made and found to be totally inactive.[4] Can one not change anything in this particular group and retain activity? How about the length? Homoarginine was used (an arginine with an addition of CH_2 in the side chain) and found not to be tolerated by relaxin-receptors. Nothing, and absolutely nothing, could be changed in this configuration, although the residues between the arginines varied significantly in various relaxins. Could inactivation of the relaxin molecule be a simple structural effect? This argument can be disregarded because all changes were made on the surface of the B chain helix using conservative replacements. In addition, CD-spectroscopy showed that the spectra of all analogs were superimposible with the human relaxin spectrum indicating structural identity.[4]

Completely, and in a way, happily rebuffed by the guanidino groups that were towering over the B chain helix in all relaxins we turned to other constant features. The glycine in position B12 was present in every relaxin whereas B24 was constant except for the hamster-derived molecule.[5] Constant residues in this mutation-happy environment of the neo-Darwinian molecular world means

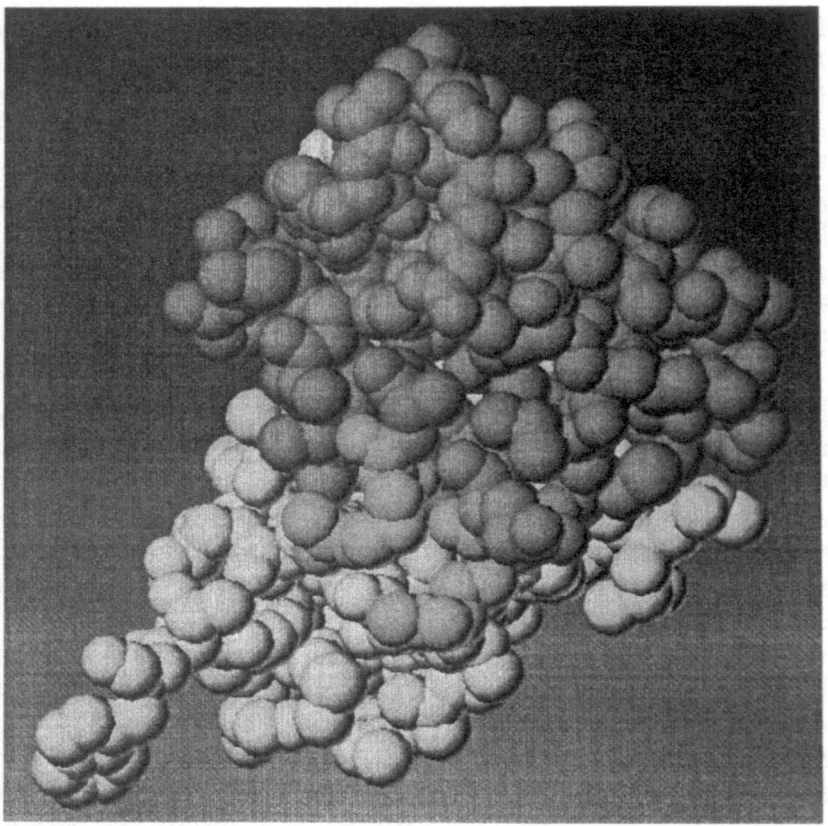

Fig. 10.3. Proposed relaxin receptor interaction site involves the two arginines on the B chain helix (The X-ray data were derived from the Brookhaven databank, where they were deposited by Eigenbrot et al 1991).

(A, above) Three-dimensional structure of the relaxin dimer. The two monomers are shown in gold and purple, respectively. The proposed active site arginines are shown from one monomer only (cyan), protruding from the B chain helix and stabilized by residues on the surface of the second monomer.

(B, opposite) Secondary structure of one monomer with arginines pointing into the water. The A chain is shown in cyan and the B chain in gold.

See color insert, pages 194-195.

that these residues were protected from changes because of functional necessity. All animals with relaxins that had suffered mutations at these points would not have survived and therefore whenever one isolates a human relaxin one sees glycine in position B12 and in B24; or so goes the story. The arginines B13 and 17 in all relaxins are an excellent example of constancy due to absolute functional requirement. Tempting as it is to conclude that therefore all other

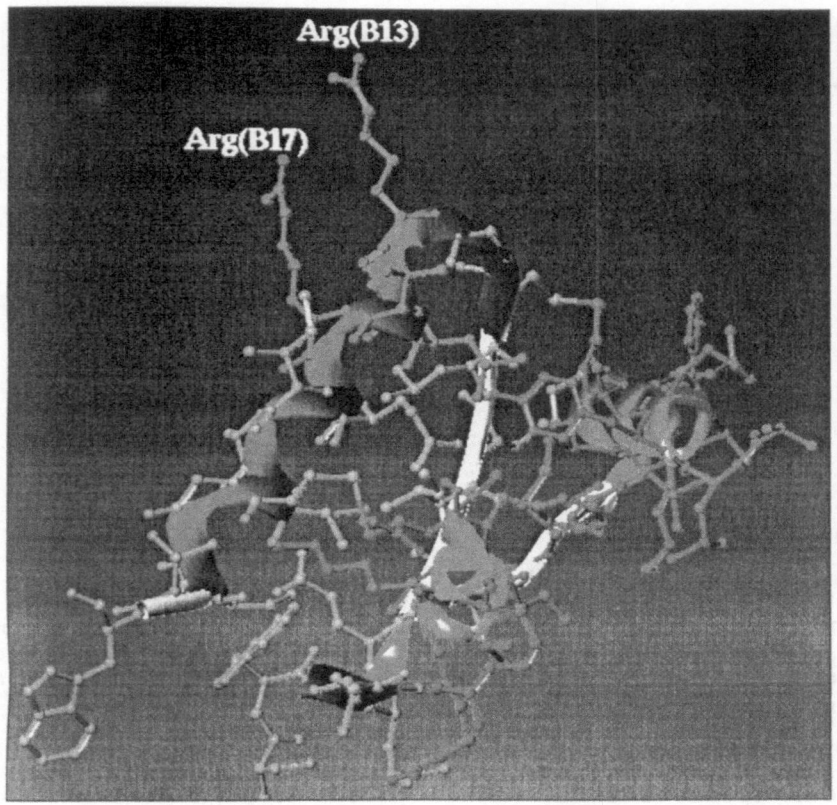

constant residues are important, this extrapolation may not be true. Innate skepticism made us prepare four B chain derivatives, replacing glycine in B12 with L-alanine or D-alanine and glycine in B24 with the same substituents. Although none of the analogs was as active as human relaxin in the mouse symphysis pubis assay (Fig. 10.4) L-alanine(B12) relaxin showed distinct activity at 1 µg/mouse and, surprisingly, D-alanine(B12) was even better in the in vivo mouse assay while in the receptor-binding assay L-alanine(B12) was as potent as human relaxin and D-alanine(B12) showed a 25% to 50% reduction in affinity in vitro.[6] Thus, the absolutely constant glycine in position B12 in human relaxin or equivalent positions in other relaxins could not only be replaced by an alanine derivative but even by the corresponding D-amino acid without significantly reducing or impairing the bioactivity of this relaxin. When comparing bioactivities it is important to consider that some intact relaxins from

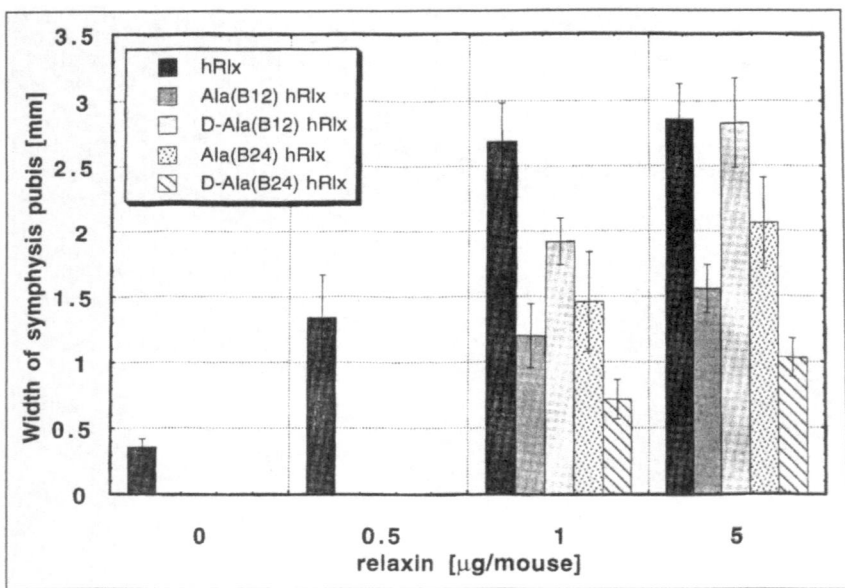

Fig. 10.4. Mouse symphysis pubis assay of human relaxin analogs with substitutions in position B12 and B24, replacing conserved glycines in these positions by either L-alanine (Ala) or D-alanine (D-Ala). The response of a negative control group is shown with a relaxin doses of 0. The bars represent standard errors.

other species where much less active than these derivatives. The B24 glycine residue could be replaced by L-alanine with only slight impairment of bioactivity and a 5- to 10-fold reduction of in vitro receptor-binding activity. The D-alanine in B24 reduced bioactivity significantly and in vitro relaxin receptor-binding by three orders of magnitude.[6] All of these activities were measured against mouse receptors as a standard assay.

The observation of constancy of residues in homologous proteins may not a priori be interpreted as a sign of structural or functional importance; the idea may be attributed to a misconception concerning the evolution of biological function. As far as our studies are concerned the arginines in the B chains in positions 13 and 17 in human relaxin and equivalent positions in other relaxins are the only constant residues that are also critically important for receptor interaction. There are ancillary conditions that increase or decrease the affinity for the receptor which are responsible for the different potencies of species-specific relaxins.

References

1. Büllesbach EE, Schwabe C. Total synthesis of human relaxin and human relaxin derivatives by solid phase peptide synthesis and site-directed chain combination. J Biol Chem 1991; 266:10754-10761.
2. Kaiser E, Colescott RL, Bossinger CD et al. Color test for detection of free terminal amino groups in the solid phase synthesis of peptides. Anal Biochem 1970; 34:595-598.
3. Sarin VK, Kent SBH, Tam JP et al. Quantitative monitoring of solid phase peptide synthesis by the ninhydrin reaction. Anal Biochem 1981; 117:147-157.
4. Büllesbach EE, Yang S, Schwabe C. The receptor-binding site of human relaxin II: A dual prong-binding mechanism. J Biol Chem 1992; 267:22957-22960.
5. McCaslin RB, Renegar RH. Determination of the prorelaxin nucleotide sequence and expression of prorelaxin messenger ribonucleic acid in the golden hamster. Biol Reprod 1995; 53:454-461.
6. Schwabe C, Büllesbach EE. Relaxin: Structures, functions, promises and non-evolution. FASEB J 1994; 8:1152-1160.

CHAPTER 11

Merely a G (Glycine)

The relaxin work lasted through several grant applications and it has become common knowledge by now that the route from one grant to another is perilous at best. Several strategic decisions are necessary in terms of a funding agency and review committees in order to improve the odds of success. For awhile we worked with the reproductive division of Child Health and Human Services and found it extremely difficult to impress this group with the importance of relaxin for their mission!? This was a conflict that led to a temporary break in our grant support. One of all the derivatives we had proposed to synthesize, in order to further explore the importance for bioactivity of the various surface regions of relaxin, involved the replacement of the absolutely constant glycine in the A chain loop, and the comment was that "replacement of a mere glycine is hardly an adventurous proposal." The reader may be wondering for a moment about this remark and why it made it to the headline of this chapter. There is no general correlation between the size of an amino acid and its relative importance in biology. Osteomalasia is a fatal disease caused by the exchange of only one of 900 glycines in type I collagen molecules for any other amino acid.

Insulin always has a large hydrophobic amino acid in that position (A10) which corresponds to the glycine A14 in human relaxin (Fig. 11.1). The X-ray crystallographers have assured us that in this case the glycine in relaxin was of no structural significance because the A chain loop points into the water and the space occupied by glycine could easily accommodate the largest amino acid in the biological repertoire. Could it really?

We changed the NIH sponsoring division, the study section, and subsequently the glycine in position A14, and replaced it by an isoleucine. The (latter) exchange produced a blue shift in the CD

Relaxin and the Fine Structure of Proteins, by Christian Schwabe and Erika E. Büllesbach. © 1998 Springer-Verlag and R.G. Landes Company.

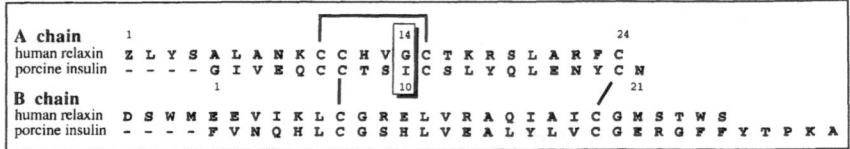

Fig. 11.1. Primary structures of human relaxin II and porcine insulin.

signal of relaxin (Fig. 11.2a), and eliminated most of the receptor-binding in vitro (Fig. 11.2b) as well as the bioactivity in the intact mouse! (Fig. 11.2c).[1] This result stressed the imagination. There was indeed space enough to accommodate isoleucine in position A14, and furthermore nothing in relaxin was farther away from the receptor-binding site than the A chain loop. Somehow the ideas did not seem to hold together and it was time to look for additional ways to test what had been observed. Looking back to insulin it was also important to find out whether isoleucine was as much a necessity for maintaining the active insulin structure as the glycine was for relaxin. We synthesized a glycine(A10) porcine insulin, made an insulin receptor preparation from human placenta, and tested the receptor-binding of glycine(A10) insulin to its own receptor. Compared to the native insulin binding of the glycine(A10) insulin to its receptor was reduced by two orders of magnitude and the conclusion, drawn on the basis of these experiments, was simply that insulin needs a large amino acid in the A10 position and relaxin needs a small one in the equivalent position A14.[1] One does not know precisely what kind of steric effect causes inactivation and how it is transmitted to the binding site but at least there are some leads.

So it happened that the smallest amino acid led to two significant observations. Not only was the importance of glycine for the active conformation of relaxin established but it also provided an upstart insulin group with a discovery that had escaped the establishment for the past 50 years. So much also for crystallographers' ability to predict the effect of specific amino acids on protein structure. The correlation is there but it is too complex for us to untangle; it is a chaos problem.

The phenomenon was interpreted as a long-range steric effect that would cause these highly similar molecules to assume a slightly different active conformation.[1] Meanwhile the X-ray structure of relaxin had been solved by Eigenbrot et al and the relaxin/insulin

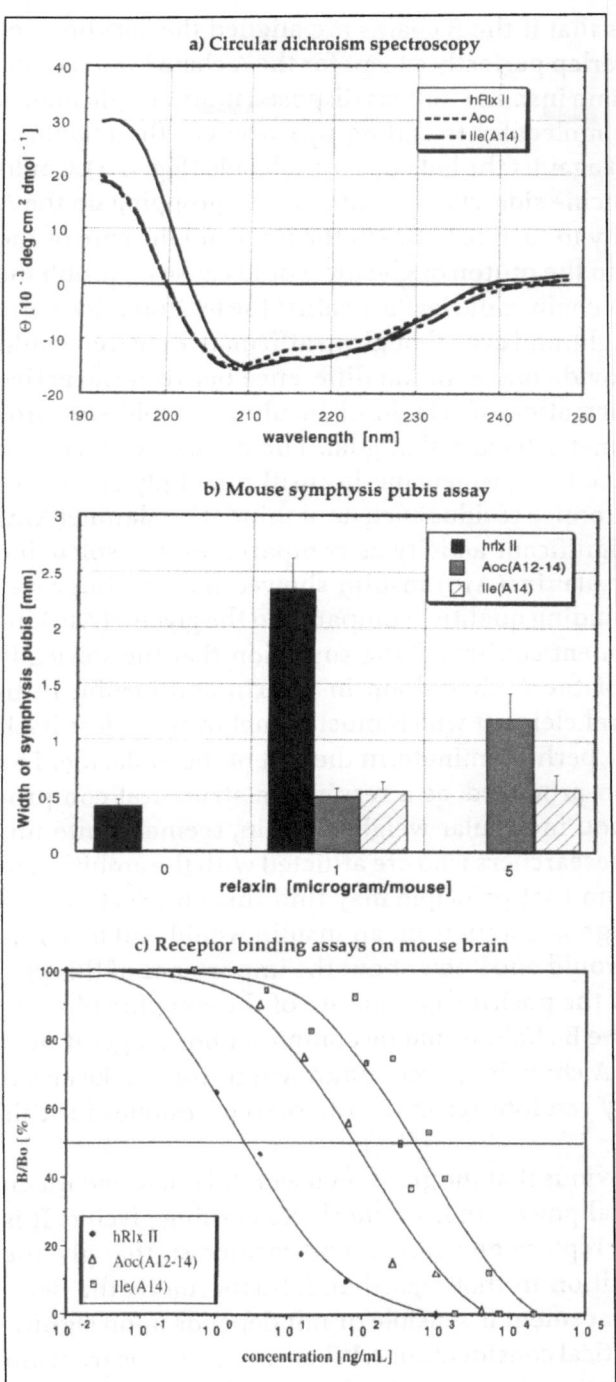

Fig. 11.2. Comparison of human relaxin with Ile(A14) relaxin and Aoc(A12-14) relaxin. Panel a, CD spectra; panel b, mouse symphysis pubis assay; panel c, receptor-binding assay.

comparison shows that if the B chains are aligned the backbone of both molecules overlap perfectly, except for the A chain loop region[2] (Fig. 11.3). The A chain insulin loop was disposed more at right angles to the core of the molecules (standing up) whereas the relaxin A chain was lying flat against the bulk of the molecule (Fig. 11.4). Could it be that the isoleucine side-chain in insulin was propping up the A chain by its inability to tuck in between the loop and the core of the molecule? In relaxin the proton of glycine would cause no problems and therefore the A chain could lie flat against the hydrophobic core. An intriguing thought, and even though an affirmative answer would not necessarily provide one with the difference between the active and inactive conformation of relaxin or insulin, it would still provide another parameter toward that goal. The obvious experiment to do was to replace both, isoleucine in insulin and glycine in relaxin, with a compromise residue such as alanine. The alanine(A14) relaxin regained significant activity as compared to the isoleucine derivative, and the alanine(A10) insulin showed marked improvement in receptor-binding qualities compared to the glycine(A10) insulin.[1] This experiment confirmed the suspicion that the sterically induced position of the A chain-loop in relaxin and insulin is an important structural element which must somehow manifest itself in a structural shift, perhaps minute, in the rest of the molecule. The allosteric transition produced, as a result of a structural compromise in a small 6000 molecular weight protein, seemed quite unusual. Thoughtful researchers who are afflicted with the ambition to design proteins from first principle may find this observation disconcerting. To design a relaxin from an insulin would still be a formidable job. How would one know about the importance of this glycine, particularly if the position and nature of the receptor-binding site in the core of the B chain would be common knowledge? If only the residues of the A chain loop were unknown it would take about 20^3 trials to find, by random selection, the correct residues for this loop.

By now it is obvious that the grant reviewer did not spend much time and intellectual power on his remark concerning glycine. It is equally difficult to replace any amino acid residue so that glycine had no special position in that regard and, furthermore, the decision whether a replacement is sensible or not depends upon significantly deeper analytical considerations. What else can one learn from this incident? The message seems to be that intuition and self-confidence are important, though irrational, ingredients for success.

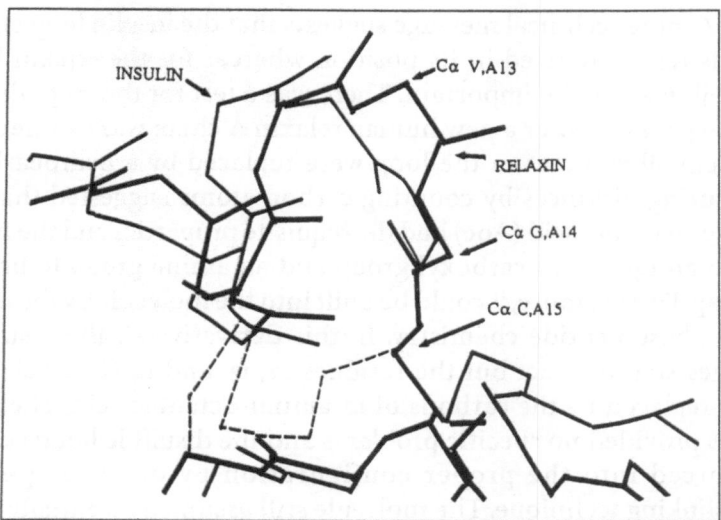

Fig. 11.3. A chain loops of insulin (thin line) and relaxin (thick line) derived from X-ray data. Reprinted with permission from: Büllesbach EE, Schwabe C. J Biol Chem 1994; 269:13124-13128.

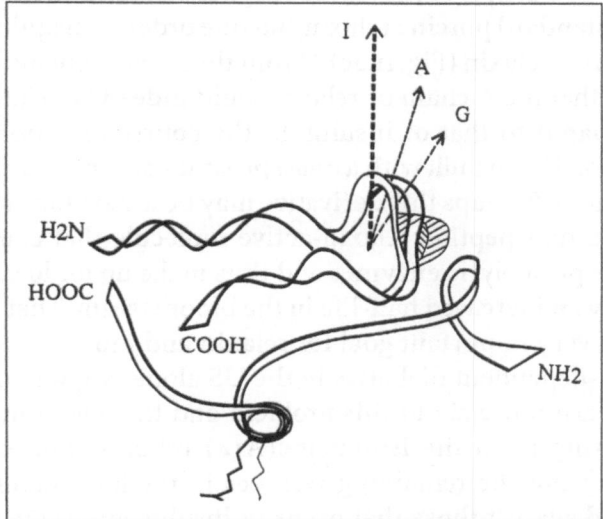

Fig. 11.4. Schematic of the insulin-relaxin loop types and the projected intermediate position of the A10/A14 alanine derivatives of both hormones. Hypothetical vectors placed against the loop planes help to visualize the differences. Reprinted with permission from: Büllesbach EE, Schwabe C. J Biol Chem 1994; 269:13124-13128.

A more technical message suggests that the insulin loop structure is relatively fixed in its position whereas for the relaxin loop flexibility would be important. There was a test for this hypothesis, and to get an answer a new human relaxin A chain was synthesized wherein all residues in the loop were replaced by a hydrocarbon. Measuring distances by counting carbon atoms suggested that ω-amino-octanoic acid (Aoc) had the requisite properties and the functional groups, i.e., a carboxyl group and an amino group to bridge the gap. Furthermore, it could be built into the molecule by the usual solid phase peptide chemistry. In this derivative all the disulfide bridges stayed intact but the residues 12, 13, and 14 (His, Val, Gly) were replaced by the carbons of ω-amino-octanoic acid. The synthesis provided no specific problems and the disulfide bonds could be forced into the proper configuration by our site-specific crosslinking technique. The molecule still assumed essentially a relaxin-like structure as identified by CD spectroscopy (Fig. 11.2a). The Aoc derivative showed distinct biological activity in mice at a dose of 5 μg/animal (Fig. 11.2b) and had remarkable receptor-binding activity at a two orders of magnitude higher concentration compared to native standard porcine relaxin, but one order of magnitude lower than Ile(A14) relaxin (Fig. 11.2c).[3] From these experiments one could conclude that the A chain of relaxin could indeed be relatively flexible compared to that of insulin. In the course of action the first molecule had been built with a quasi prosthesis, namely a non-amide carbon chain. Perhaps this derivative may be a start toward the synthesis of a non-peptide relaxin-active molecule that could escape the body's proteolytic enzymes and thus make up for lower specific activity by an increased half-life in the bloodstream. That would indeed be a very important goal for relaxin and more so for millions of insulin-dependent diabetics in the US alone. Anyway, one would have to learn more about this problem and the one of many questions arising from the isoleucine(A14) relaxin experiment was whether or not the remaining residues in the loop should also be exchanged against those that occur in insulin, and if the total loop exchange may yield an active derivative. Relaxin was synthesized with all A chain loop residues exchanged for those in insulin leading to a Thr-Ser-Ile (A12-14) human relaxin II. This was one of the most inactive A chain loop relaxin derivatives (Table 11.1) so that the answer stood out quite clearly, i.e., the glycine in position A14 is very important because of smallness, the remaining residues in the loop

Table 11.1. Biological activity and receptor-binding activity of human relaxin derivatives with modifications in the A chain loop

Relaxin Derivative	Mouse Symphysis Pubis Assay	Receptor Binding ED_{50}
human relaxin II	+++	1 to 6 ng/ml
L-Ala(A14)	++	10 to 15 ng/ml
D-Ala(A14)	+	50 ng/ml
Ile(A14)	−	300 to 800 ng/ml
Aoc(A12-14)	+	50 to 110 ng/ml
Thr-Ser-Ile(A12-14)	−	350 ng/ml
Ser(A10)Ser(A15)	−	1000 ng/ml
Ala(A10)Ala(A15)	+	50 ng/ml

+++ fully active at 1 µg/mouse; ++ low activity at 1 µg/mouse; + low activity at 5 µg/mouse; − no activity at 5 µg/mouse

make no difference or cause perturbations, but the replacement of all residues by a flexible hydrocarbon chain still allows for significant relaxin activity.[3]

At this point the reader may be wondering whether the loop is required at all, and in line with that idea cysteine A11 and cysteine A15 were replaced with serine residues. This was too much; the biological activity disappeared with this change. Still there were two possible explanations, namely (Table 11.1) that serine is the wrong residue to substitute for cysteine in spite of the fact that it is commonly used, or that the loop per se is required. Serine may be too large and the polar side-chain cannot be comfortably tucked away into the space that was occupied by the hydrophobic sulfur. Only an experiment could decide between protein chemists' paranoia and a justified concern, and consequently another derivative was synthesized wherein the cysteines in A11 and 15 were exchanged for alanines. These derivatives showed a slight improvement over the serine derivative. At least one can say that the substituents used to replace cysteine make a difference and that alanine is better than serine for that purpose in relaxin.[3]

This was the level of understanding of the function of the A chain loop in relaxin until, in 1997, Parry et al published the cDNA sequence of marsupial relaxin, the first natural relaxin in which glycine(A14) is replaced by serine.[4] It was of interest to know whether or not the "essential glycine" interpretation requires revision and consequently this molecule was synthesized and tested. Compared to human relaxin, marsupial relaxin had approximately 1% activity in the in vitro receptor-binding assay. Since marsupial relaxin has only 47% homology to human relaxin there remain questions as to how much of the lost activity is attributable to the serine in position A10. Synthesis of an A14 glycine derivative of marsupial relaxin will provide the answer.

Much had been learned about the relaxin structure and function relationship, including the fact that interpretation of the results obtained by replacement is a much more complex problem than a literature survey would lead one to believe. Conversely, if what had been learned is of any value we should be able to make predictions and to obtain experimental verification. That is certainly still true but the experience described here should infuse a substantial amount of caution into interpretive aspects of protein structure and function work.

Similar to the situation of glycine A14 is the constant appearance of a pair of basic amino acid residues on the C-terminal A chain helix, with skate relaxin as the only exemption. In spite of earlier work on the structure and function relationship the consistent appearance of basic amino acid residues in the A chain still inspires relaxin researchers to discuss this feature as potential relaxin receptor-binding site. True, the original scanning through the various regions of the relaxin surface only suggested that the lysine residues were not important but did not provide conclusive evidence. To clarify the situation it was decided to re-examine that problem in an orderly fashion. A human relaxin A chain synthesized with two arginines (A18, A22) and two lysines (A9, A17) replaced by citrulline, yielded upon combination with a normal relaxin B chain a well-behaved soluble molecule with an isoelectric point in the acidic range close to that of insulin. By molecular modeling, using the coordinates published by Eigenbrot et al,[2] one could highlight the basic residues in the A chain and observe that they all pointed away from the receptor-binding site-bearing B chain (Fig. 11.5). Tetra-citrulline relaxin showed no impairment of biological activity or receptor-binding in vitro.[5]

From tetra-citrulline relaxin one could learn that the basic residues in the A chain could be used for specific labeling with radiolabeled Bolton Hunter reagent[6] or with biotin[7] for avidin selection technology. There was hope that tetra-citrulline relaxin could, for

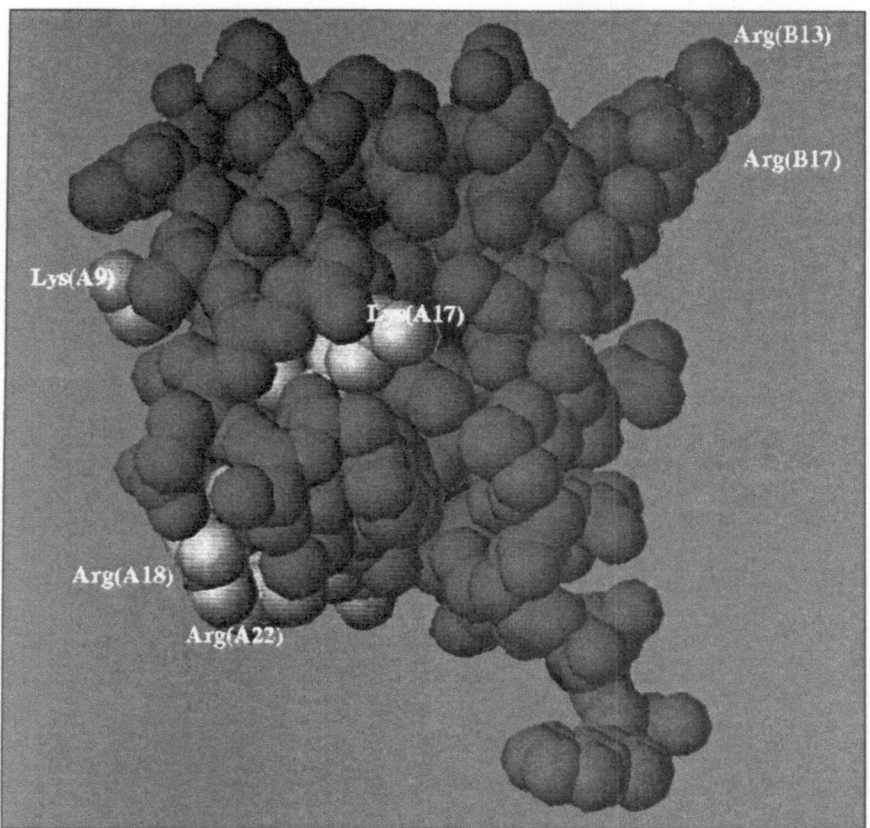

Fig. 11.5. X-ray structure of relaxin showing the location of the four basic residues of the A chain (yellow) in relation to the binding-site arginines on the B chain (magenta). The figure indicates that basic residues of the A chain are located on the surfaces opposite to the active site arginines. See color insert, page 195.

instance, solve the problem of high nonspecific binding of relaxins to relaxin receptor-bearing tissues which had been attributed to the high density of basic residues in the A chains. Had this been the case, replacement of the positively charged side chains by the uncharged urea group should have eliminated that problem. It did not. The nonspecific binding of a tracer made from tetra-citrulline relaxin was still fairly high compared to the specific binding so that our gain from these experiments was essentially limited to the realization that the so-called conserved basic region appears to be conserved for no obvious reason.

While the tetra-citrulline A chain relaxin did not give us a new tracer or any other practical advantage, the molecule made history by becoming the first relaxin ever to appear on a journal cover (*Endocrine Journal* Vol.1, 1993).

References

1. Büllesbach EE, Schwabe C. The functional importance of the A chain loop in relaxin and insulin. J Biol Chem 1994; 269:13124-13128.
2. Eigenbrot C, Randal M, Quan C et al. X-ray structure of human relaxin at 1.5 Å: Comparison to insulin and implications for receptor binding determinants. J Mol Biol 1991; 221:15-21.
3. Büllesbach EE, Schwabe C. Structural contribution of the A chain loop in relaxin. Int J Peptide Protein Res 1995; 46:238-243.
4. Parry LJ, Rust W, Ivell R. Marsupial relaxin—complementary deoxyribonucleic acid sequence and gene expression in the female and male tammar wallaby, *Macropus eugenii*. Biol Reprod 1997; 57:119-127.
5. Rembiesa B, Bracey R, Büllesbach EE et al. The conserved basic residues in the A relaxin chains are not required for biological activity—synthesis and properties of a tetra-citrulline human relaxin II. Endocrine J 1993; 1:263-268.
6. Yang S, Rembiesa B, Büllesbach EE et al. Relaxin receptors in mice: Demonstration of ligand binding in symphyseal tissues and uterine membrane. Endocrinology 1992; 130:179-185.
7. Büllesbach EE, Schwabe C. Monobiotinylated relaxins: preparation and chemical properties of the mono(biotinyl-ε-aminohexanoyl) porcine relaxin. Int J Peptide Protein Res 1990; 35:416-423.

CHAPTER 12

The Development of Insulaxin, The First True "Zwitterhormon"

Floating on top of the world, a small world that is, one is given to making heady predictions. From what had been learned about insulin and relaxin it seemed justified to predict that, by careful compromise between relaxin-and insulin-specific requirements, a true dual hormone could be designed which would recognize the relaxin as well as the insulin receptor.

This had been tried before by random combination of insulin A and relaxin B chains and vice versa (Tregear, GW part of the discussion in ref. 1). These constructs exhibited no relaxin activity. There was, however, no absolute proof that the chain combination had worked properly so that the experiment could not be interpreted without reservation. Today we know that functional hybrids could not have been produced by that method. The attentive reader will have no difficulties identifying the problem since the receptor-interaction site located on the B chain acquires the active conformation only when the A chain has a flexible, relaxin-type A-chain loop. As this chapter develops the thoughts will be reiterated in greater detail that went into the design of dual activity into one primary sequence.

Human nature would have made us do these experiments regardless of utility but, as it happens, evidence had appeared in the past that indicated a role for relaxin in combination with insulin during pregnancy diabetes.[2,3] Completely unknown biological processes sometimes cause insulin receptors in non-diabetic expectant mothers to become insensitive to insulin. Evidence obtained from experiments with pregnant rat epididymal fat cells suggests that relaxin is able to induce the insulin receptor to regain affinity for its natural ligand, presumably by an intramembranous interaction

Relaxin and the Fine Structure of Proteins, by Christian Schwabe and Erika E. Büllesbach. © 1998 Springer-Verlag and R.G. Landes Company.

known as "receptor crosstalk".[2,3] Pregnancy-induced diabetes therefore should at least in theory be treatable with a mixture of insulin and relaxin rather than insulin alone of which there is no deficiency. A hormone with insulin and relaxin activity combined in known proportions could again in theory be very useful to manage pregnancy diabetes because there would be no dosing problem as regards the relative amounts of insulin and relaxin activity given, but also because of the relatively low intrinsic activity of insulaxin.

The reader will recall that the two arginines in the B chain of relaxin are absolutely required for receptor-binding and should be preserved in insulaxin. On the other hand the core of the insulin B chain was not suspected to be part of the insulin receptor-binding region[4] and could therefore be adapted to fit the relaxin receptor. Lining up the insulin B chain with a human relaxin B chain at the cysteine residues (Fig. 11.1), it is noticed that serine and glutamic acid needed to be replaced with arginines such as to match the relaxin receptor-binding site. A series of drawings (Table 12.1) will orient the reader with respect to sequential substitutions made in the insulin molecule in order to make it into a relaxin receptor ligand. The first derivative was called RR-insulin (remembering that arginine is abbreviated by a capital R in the single letter code for L-amino acids). RR-insulin did not bind to the mouse relaxin receptor.

It is necessary to sweep back a little to bring the reader up on an important collateral discovery concerning an antibody known in the literature as R6. This polyclonal rabbit antiporcine relaxin antibody crossreacted with many of the relaxins from different species, i.e., human,[5] dog,[6] Syrian hamster.[7] High species variability of relaxin suggested that the only common motif may be the receptor-binding site and it was suggested that binding of relaxin to either the antibody R6 or the receptor has similar, if not identical, structural requirements.[8] Insulin was not recognized by this antibody,[9] but RR-insulin had become an antigen with 25% of the R6/relaxin affinity.[8] This was the strongest support yet for the suggestion that the two arginines in the relaxin B chain are indeed the sites for relaxin receptor interaction. The important aspects about the R6 recognition site is that almost all residues that showed no relaxin activity either in vivo or in vitro against the relaxin receptor membrane preparation also did not interact with the R6. Modifications of the important B chain residues nearly completely eliminated the R6 interaction in relaxins whereas the A chain changes only reduced R6

Table 12.1. *The systematic development of insulaxin, an insulin analog, with crossreactivity to the insulin- and relaxin receptor*

Insulin Analog	Anitbody R6	Insulin Activity (ED50)
Insulin	0	100%
RR-insulin	25%	2.3%
GRR-insulin	33%	0.4%
GRER-insulin	98%	7.6%
GRER-dpp-insulin	420%	10%
Human relaxin II	100%	0

Antibody R6 is a polyclonal anti-porcine relaxin antibody which mimics in part the active site of relaxin. Insulin activity was determined on crude membrane preparations from human placenta.

recognition.[8] These studies strongly suggested that R6 was indeed directed against the relaxin receptor interaction site and that the antibody would be a useful tool to assay for a relaxin activity of the insulin-relaxin constructs.

Next Ile(A10) was replaced by the (almost) invariable glycine of the relaxin A chain loop to generate GRR-insulin. This insulin/relaxin hybrid did not change affinity to R6 significantly when compared to RR-insulin, did not crossreact with the relaxin receptor, but reduced the affinity to the insulin receptor to 0.4% relative to native insulin. GRR-insulin in which the three crucial amino acids of relaxin were introduced was a molecule without hormonal function! The two arginine residues on the B chain and the glycine on the A chain are necessary but not sufficient for binding and therefore it was necessary to explore other residues that could adjust the insulin derivative to an active relaxin molecule.

The region between the two arginines in human relaxin II and porcine relaxin was occupied by glutamic acid, leucine and valine. We assumed that glutamic acid might lock a neighboring arginine side chain into a different position via a salt bridge and thereby generate a more active analog. The position corresponding to glutamic acid is a histidine in insulin and since it had been observed that the replacement of histidine by either aspartic acid or glutamic acid could resurrect insulin analogs of low potency[10] the next plausible step was to replace histidine with a glutamic acid in the B10 position to generate GRER-insulin.

Every time the term replacement is used in this context it means a new synthesis of a chain and a new combination with the appropriate partner-chain, but with the experience accumulated, it took only about a month to prepare a new derivative. That was a small price to pay for the absolute certainty that nothing unintended had been changed. GRER-insulin showed a dramatic jump in affinity for the R6 antibody which was most encouraging. Preliminary bioassay work at this point showed that the construct was not active in widening the symphysis pubis in estrogen-primed mice, and did not crossreact with relaxin receptors on crude membranes from mouse brain, but that the insulin activity had improved. The insulin-likeness of our molecule could be further improved by changing the length of the B chain. The insulin B chain C terminus is longer by five amino acid residues than the relaxin B chain. This change called for the synthesis of an insulin B chain which lacked the sequence B26-30 (Y-T-P-K-A) known as despentapeptide insulin which is amidated at the new C-terminal phenylalanine. Despentapeptide-insulin-amide retained the full insulin activity[11] and it seemed therefore a good idea to shorten the hybrid to match the relaxin B chain length (Fig. 11.1). The analog, named GRER-dpp-insulin (dpp = despentapeptide), bound the anti-porcine relaxin antibody R6 as well as the original antigen (porcine relaxin) and better than native human relaxin!

For reasons that we do not yet understand the mouse relaxin receptor barely, if at all, recognized our decoy. The truncated insulin

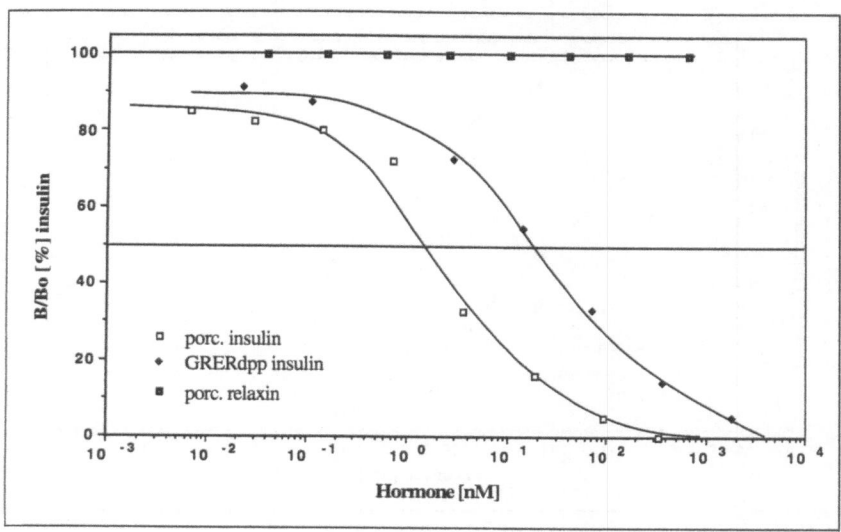

Fig. 12.1. Receptor-binding assay of GRER-dpp-insulin, insulin, and relaxin to crude membrane preparations of human placenta. Porcine [^{125}I-Tyr(A14)] insulin was used as tracer. Reprinted with permission from: Büllesbach EE, Steinetz BG, Schwabe C. Biochemistry, 1996; 35:9754-9760. Copyright 1996 American Chemical Society.

into which Gly (A10), Arg (B9), Glu (B10) and Arg (B13) had been introduced (GRER-dpp-insulin) reacted specifically and reversibly with the human placental insulin receptor (Fig. 12.1). Somehow it seemed impossible that the relaxin receptor would not recognize our construct.

The mouse receptor, although generally promiscuous toward other relaxins, might for some reason not be the correct one to test. Therefore rat brain receptor membranes were prepared and the experiments repeated with GRER-dpp-insulin. The results displayed in Figure 12.2 gave rise to another celebration. Yes, we had a dual hormone, a genuine Zwitterhormon, that displayed both insulin and relaxin receptor-binding in one primary sequence, which meant that the new concepts about the structure-function relationship for these two hormones are quite reasonable.[8] Of course, some fine adjustments will yet be necessary to obtain this dual activity toward the human receptors so that the clinical researchers might put this compound through the paces and evaluate its utility for humans, in particular for pregnancy-induced diabetes.

The two arginines had been largely accepted as receptor-binding features, but this study serendipitously added further support to

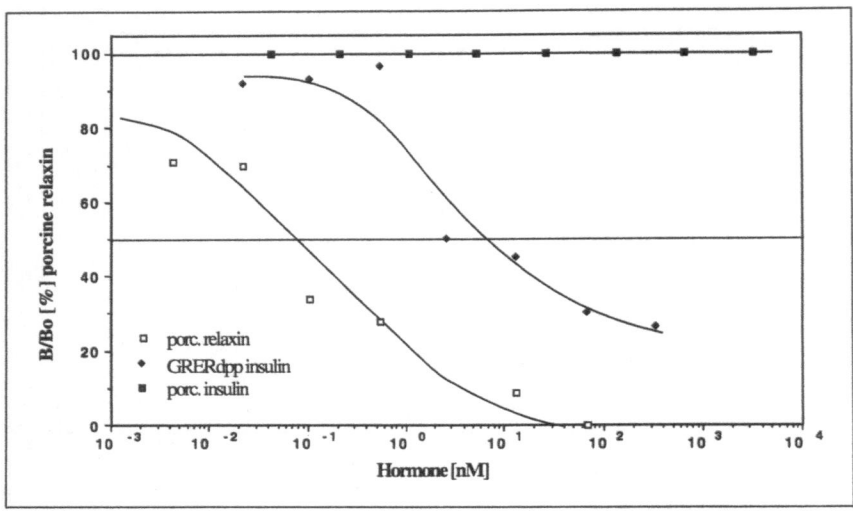

Fig. 12.2. Receptor-binding assay of GRER-dpp-insulin, insulin and relaxin to crude membrane preparations of rat brain. [^{125}I]3,5-diiododesaminotyrosyl(A1) porcine relaxin was used as tracer. Reprinted with permission from: Büllesbach EE, Steinetz BG, Schwabe C. Biochemistry 1996; 35:9754-9760. Copyright 1996 American Chemical Society.

that idea and secured at the same time strong evidence for the concept of long-range steric effects in a small protein hormone. Still it remains difficult to imagine how the A-chain portion of relaxin could disturb a relatively stable helix sufficiently to influence receptor interaction. Surely the reader would demand another installment of this story a few years down the road.

The student in particular will notice that conceptual advances realized through this work were obtained with relatively standard methods of peptide chemistry and that the methods of molecular biology could not have provided this information. Citrulline and D-amino acid substitutions, for example, cannot be done that way and yet it may be a learning experience to guess why this information had its problems on the way to the public. It was too technical (one recommended a methods journal!) and not elegant (meaning: no gene technology). Not to despair students of science, it will all straighten out in time!

What really has been done that would warrant attention? Different proteins have been covalently linked before and the complex retained both activities, and chimera have been produced by cloning techniques which combined an external receptor domain with the internal portion of another receptor. Insulaxin is different in that the same primary sequence has been tweaked into performing two totally unconnected bioactivities and simultaneously to impart relaxin antibody recognition into an insulin, not by the modern random approach (20×10^{50} possibilities) but rather by targeted substitution of only 4 residues, and that can fill one with great satisfaction.

References

1. Isaacs N, Dodson G. Model of relaxin. In: Bryant-Greenwood GD, Niall HD, Greenwood FC, eds. Relaxin. North Holland: Elsevier, 1981:101-109.
2. Olefsky JM, Saekow M, Kroc RL. Potentiation of insulin binding and insulin action by purified porcine relaxin. Ann NY Acad Sci 1982; 380:200-216.
3. Jarrett JC, Ballejo G, Saleem TH et al. The effect of prolactin and relaxin on insulin binding by adipocytes from pregnant women. Am J Obst Gynecol 1984; 149:250-255.
4. Schäffer L. A model for insulin binding to the insulin receptor. Eur J Biochem 1994; 221:1127-1132.
5. O'Byrne EM, Carriere BT, Sorensen L et al. Plasma immunoreactive relaxin levels in pregnant and nonpregnant women. J Clin Endocrinol Metabol 1978; 47:1106-1110.
6. Steinetz BG, Goldsmith LT, Lust G. Plasma relaxin in pregnant and lactating dogs. Biol Reprod 1987; 37:719-725.
7. Steinetz BG, O'Byrne EM, Goldsmith LT et al. The source of relaxin in pregnant Syrian hamsters. Endocrinology 1988; 122:795-798.
8. Büllesbach EE, Steinetz BG, Schwabe C. Chemical synthesis of a Zwitterhormon, insulaxin, and of a relaxin-like bombyxin derivative. Biochemistry 1996; 35:9754-9760.
9. O'Byrne EM, Steinetz BG. Radioimmunoassay (RIA) of relaxin in sera of various species using an antiserum to porcine relaxin. Proc Soc Exp Biol Med 1976; 152:272-276.
10. Burke GT, Hu SQ, Ohta N et al. Superactive insulins. Biochem Biophys Res Commun 1990; 173:982-987.
11. Fischer WH, Saunders D, Brandenburg D et al. A shortened insulin with full in vitro potency. Biol Chem Hoppe Seyler 1985; 366:521-525.

CHAPTER 13

A Surprising Message from Murines

Rat relaxin had been known for about 18 years and had been thoroughly characterized mainly through by Sherwood et al.[1-4] The relaxin used in those studies had been isolated by the method of Walsh et al in which homogenized ovaries were extracted with a mixture of trifluoroacetic acid (15%), formic acid (5%), sodium chloride (1%) and hydrochloric acid (1M).[5,6] The conditions seemed rather rough but an active hormone was obtained which was used for all subsequent studies until very recently. Rat relaxin (Fig. 13.1), always considered to be a relaxin of low potency, made a natural connection to our structure function studies, in particular the beneficial effect of replacement of histidine B10 in insulaxin for a glutamic acid. In the rat this position was occupied by glycine and based on a central dogma in protein chemistry, this was an unfavorable situation. Glycine is considered a helix-breaking residue and its presence in the major B chain helix right at the receptor interaction site seemed a clear indication that glycine B14 was compromising the bioactivity of rat relaxin. The opportunity was irresistible to test one's ability to predict how certain features in a molecule would influence its bioactivity and subsequently to improve on a natural product. First rat relaxin was needed to establish a baseline and consequently a few milligrams were synthesized by the new method.[7] Surprisingly, when this material was subjected to a quantitative receptor-binding assay it proved to be as active as synthetic human relaxin II and native porcine relaxin which to date are the most active members of this family. Something seemed wrong, but after two or three repetitions rat relaxin remained just as good as porcine and human relaxin II. Eventually some isolated rat relaxin was donated by our colleague and ran for comparison with synthetic rat relaxin

Relaxin and the Fine Structure of Proteins, by Christian Schwabe and Erika E. Büllesbach. © 1998 Springer-Verlag and R.G. Landes Company.

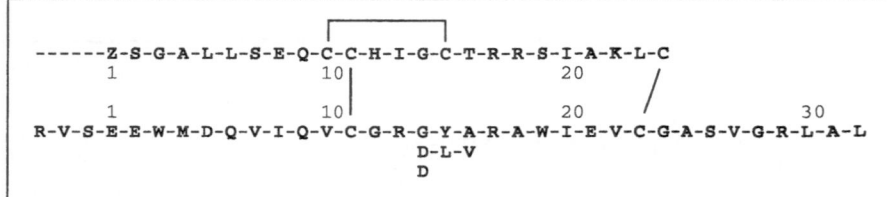

Fig. 13.1. Primary structure of rat relaxin. The human relaxin numbering system is used. Modified regions of rat relaxin analogs are indicated.

in the receptor-binding assay. Sure enough the isolated relaxin was not nearly as active as the synthetic material and the explanation appeared to be that the isolation procedure, as described at the beginning of this chapter, was too rough. As a part of the ceremonial sample exchange some synthetic rat relaxin was returned to the donor of the isolated relaxin and to date the claim of higher specific activity for synthetic rat relaxin has not been disputed. This is one of the cases where many years of painstaking work by an excellent investigator have unintentionally been torpedoed by an inadequate isolation procedure. There is something to be said for synthetic hormones if that route is available and even though synthesis comes with a price tag; eight years worth of intense studies of the physiology of a hormone do come with an incomparably higher one and should by all means be done with the appropriate reagents.

It will be noted that this outcome of the preliminary study had eliminated the reason for the suspicion that glycine in the B chain helix was indeed a perturbing agent. But then there is always hope that one could produce a hyperactive hormone and that hope led to the synthesis of a rat relaxin with an acidic residue instead of glycine in position B14. For good measure a second derivative was made wherein all three residues between the two critical arginines were replaced by amino acids that are most frequently observed in those positions in relaxins such as leucine B15 and valine in B16. The two derivatives and rat relaxin were examined extensively by CD spectroscopy and found to confirm the speculations that the helix would improve with glycine replaced by aspartic acid and that it would improve more with all three residues (GYA) replaced by DLV.[7] The receptor-binding assay on a crude membrane preparation of mouse brain showed that aspartic acid in B14 reduced the affinity for the relaxin receptor while the replacement of three amino acids (DLV relaxin) increased the affinity over that of synthetic rat relaxin.[7] The general wisdom of protein chemistry was holding up as our experiments were clearly showing that glycine in a helix region is not as good as anything else and that in one case (DLV rat relaxin) but not in the other (Asp(B14) rat relaxin) an improvement of helicity was enough to produce a slight improvement in receptor-binding. Experiments with the antibody R6 showed the same tendency. The Asp(B14) rat relaxin bound with significantly lower affinity than native relaxin with its disturbed B chain helix. When all three residues, however, were exchanged as in the DLV relaxin the binding improved dramatically, showing an affinity similar to that of human relaxin.[7] These results contradict the statement made in the previous chapter where aspartic acid in position B14 disturbed binding to the R6 antibody. To date there is no good explanation for this phenomenon but at that time we opted for a bioassay, the ultimate test for relaxin activity, to decide the issue. Figure 13.2 compares the results of running human and rat relaxin side by side in the mouse bioassay system and Figure 13.3 depicts the comparison of rat relaxin, Asp(B14) rat relaxin and DLV(B14-16) rat relaxin. The result was so startling that this experiment was repeated four times in monthly intervals. Finally it dawned on us that another truism of protein chemistry had lost its luster. By eliminating a helix-breaking residue in this active site the rat relaxin structure had been improved, the receptor-binding had been improved at least in one example, but the bioactivity had been significantly impaired. That is not to say that structure function relations were invalid but rather that they had just become more complex, more exiting, and even less predictable.

What really had happened that warrants excitement? It is not a new factor, a new enzyme or promoter, not even a new control mechanism, in fact nothing terribly obvious had occurred. Upon close scrutiny the concept of energy minimum, the guiding force for the three-dimensional organization of proteins had been uncoupled from the bioactivity optimization process. While this observation has interesting consequences for protein structure/function work, it has catastrophic ones for the hypothesis of molecular evolution. These concepts will be picked up again in chapter 19 on the impact of relaxin on evolutionary theory.

The rat relaxin story induced the synthesis of another human relaxin derivative bearing a glycine in position B14 instead of the natural glutamic acid, and the hope was to have created a superactive

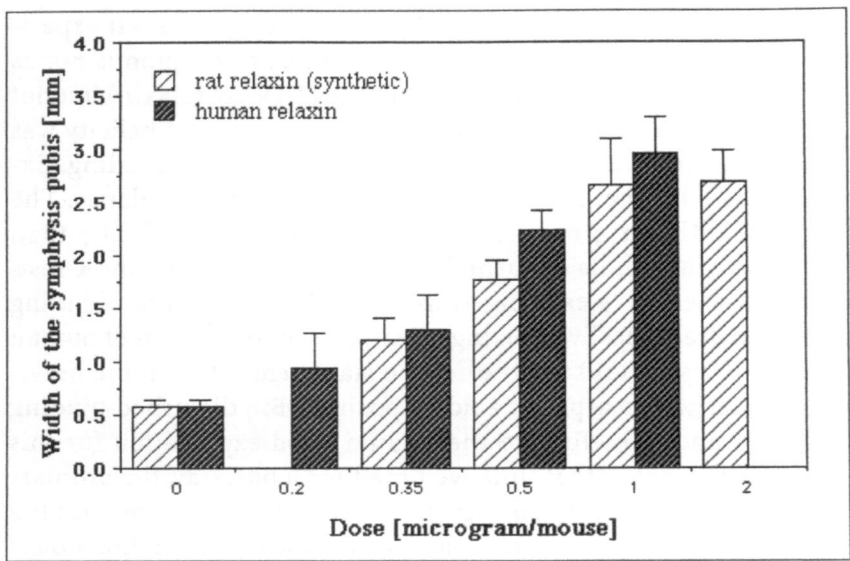

Fig. 13.2. Mouse symphysis pubis assay of synthetic human relaxin II and rat relaxin. The bars represent standard errors. Reprinted with permission from Büllesbach EE, Schwabe C. Eur J Biochem 1996; 241:533-537.

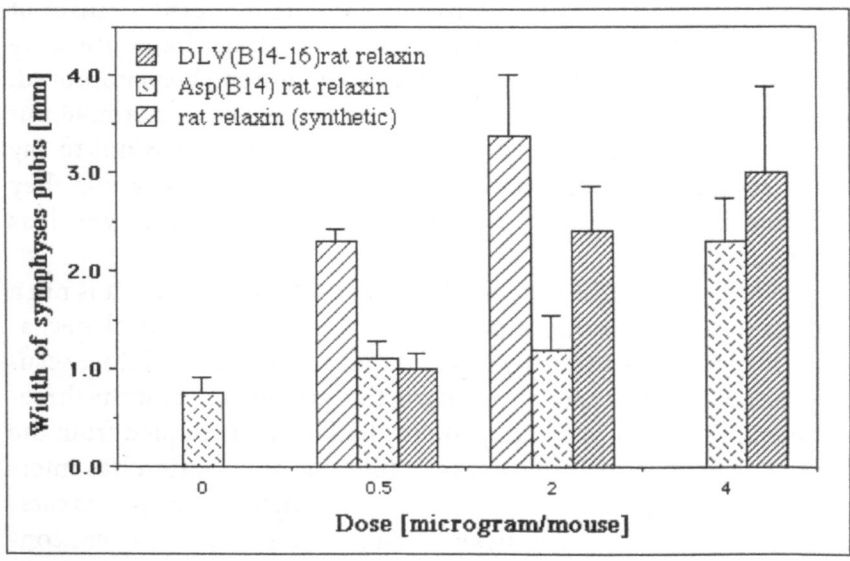

Fig. 13.3. Mouse symphysis pubis assay of rat relaxin and rat relaxin analogs. The bars represent standard errors. Reprinted with permission from Büllesbach EE, Schwabe C. Eur J Biochem 1996; 241:533-537.

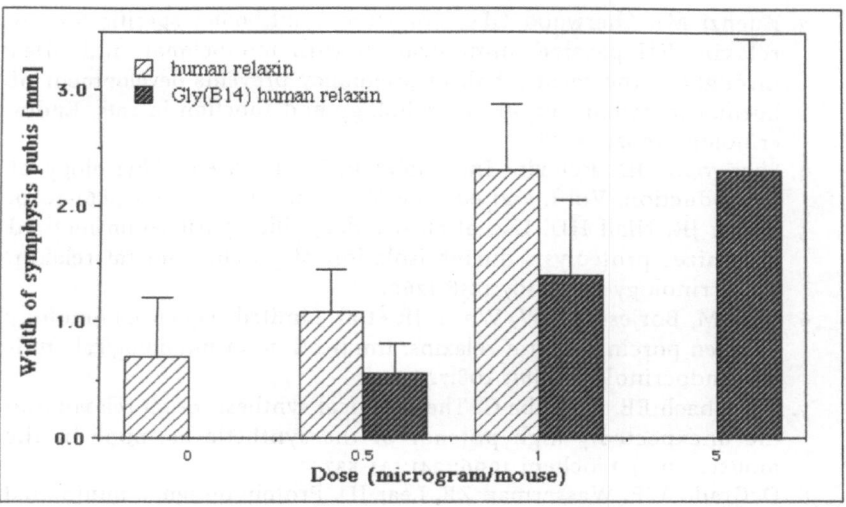

Fig. 13.4. Mouse symphysis pubis assay of human relaxin and Gly(B14) human relaxin. The bars represent standard errors.

human relaxin derivative. The synthesis went well. The chains were forced together in the correct configuration, but the Gly(B14) human relaxin bound the receptor in vitro almost as well as human relaxin. Taking little for granted it was decided to test the Gly(B14) human relaxin in the mouse symphysis pubis assay. The results depicted in Figure 13.4 show quite clearly that Gly in position B14 has reduced out bioactivity! Rat relaxin requires Gly in the equivalent of position B14 in human relaxin wherein this substitution is harmful.

If this book is about structural concepts in small proteins, right now it seems that there are none!? Not so, the problem is merely a little deeper than we thought and that is par for science. This observation tells us that, while the design from principle and experience of a fairly stable, simple protein structure may be accomplished some day,[8-10] the design of bioactivity from concepts (as opposed to trial and error) is not possible.

References

1. Sherwood OD. Purification and characterization of rat relaxin. Endocrinology 1979; 104:886-892.
2. Lao Guico-Lamm M, Sherwood OD. Monoclonal antibodies specific for rat relaxin: II passive immunization with monoclonal antibodies throughout the second half of pregnancy disrupts birth in intact rats. Endocrinology 1988; 123:2479-2485.

3. Kuenzi MJ, Sherwood OD. Monoclonal antibodies specific for rat relaxin: VII passive immunization with monoclonal antibodies throughout the second half of pregnancy prevents development of normal mammary nipple morphology and function in rats. Endocrinology 1992; 131:1841-1847.

4. Sherwood OD. Relaxin. In: Knobil E, Neill JD, eds. Physiology of Reproduction. Vol I. 2nd ed. New York: Raven Press, 1994:861-1010.

5. Walsh JR, Niall HD. Use of an octadecylsilica purification method minimizes proteolysis during isolation of porcine and rat relaxin. Endocrinology 1980; 107:1258-1260.

6. John M, Borjesson BW, Walsh JR et al. Limited sequence homology between porcine and rat relaxins: Implication for physiological studies. Endocrinology 1981; 108:726-729.

7. Büllesbach EE, Schwabe C. The chemical synthesis of rat relaxin and the unexpectedly high potency of the synthetic hormone in the mouse. Eur J Biochem 1996; 241:533-537.

8. DeGrado WF, Wasserman ZR, Lear JD. Protein design, a minimalist approach. Science 1989; 243:622-628.

9. Hecht MH, Richardson JS, Richardson DC et al. De Novo design, expression, and characterization of Felix: A four-helix bundle protein of native-like sequence. Science 1990; 249:884-891.

10. Mattos C, Ringe D. Locating and characterizing binding sites on proteins [Review]. Nature Biotechnology 1996; 14:595-599.

CHAPTER 14

The Mouse, a Very Small Rat, a Very Big Mistake

The second prominent member of the murine family, and equally as successful as its larger family member, is the mouse. Against the background of contemporary evolutionary philosophy it is easy to see the mouse as a small rat or vice versa and on that basis alone one would expect the relaxins between the two species to be nearly identical. There is always an experiment to ruin a beautiful concept. In this case it was the publication of a mouse cDNA that encoded a relaxin with the most drastic difference yet observed, i.e., a different distribution of the disulfide bonding pattern which had become the hallmark of the relaxin/insulin family.[1] While there was a fair degree of homology between other regions of the two molecules, the C-terminal cysteine had been moved out one residue relative to other relaxins. This extended the interchain disulfide loop by one extra tyrosine residue.

Was it really a relaxin? The cDNA, of course, can give only information about the primary structure and that with some uncertainty as regards the beginning and end of the A and B chains. There was enough information, however, to synthesize mouse relaxin and to measure the receptor-binding ability in a mouse assay system in parallel with porcine and human relaxin.[2] Figure 14.1 depicts mouse relaxin next to rat relaxin which has a normal disulfide crosslinking pattern. One could speculate as to what effect this extra residue would have on the normally helical portion of the C-terminal end of the A chain, but crowding to cause a loop-out and thereby interruption of the helix seemed inevitable.

The selective and sequential disulfide bond formation made it possible to produce this relaxin chemically without the nagging doubts concerning proper crosslinking. The molecule was very

Relaxin and the Fine Structure of Proteins, by Christian Schwabe and Erika E. Büllesbach. © 1998 Springer-Verlag and R.G. Landes Company.

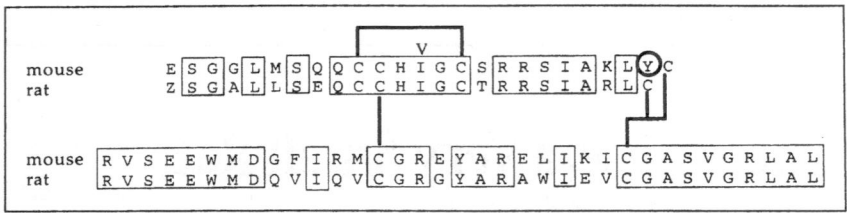

Fig. 14.1. Comparison of the primary structures of mouse relaxin and rat relaxin.

soluble in aqueous media and the CD spectrum did not show the slightest hint of helix disturbance.[2] The disulfide bond, which has a fairly rigid geometry, still could rotate enough around the carbon-sulfur bond to change direction like a weather vane, such as to accommodate the extra residue. That was a learning experience, but even more surprising was the relatively low affinity of mouse relaxin for its own receptors in spite of the fact that under physiological conditions (pregnancy) mice show a strong relaxin response[3] and serum levels of relaxin are not exceedingly high (30 ng/ml estimated from refs. 4,5). There was no hidden gene with the proper crosslink pattern in the mouse genome. A second relaxin gene in the mouse had the same type of disulfide linkages but a Val/Ile exchange in position A13[1] which was of no consequence, neither structurally nor biologically. The obvious question at this point was whether or not the extra residue in the A chain was advantageous. Consequently a mouse relaxin was synthesized without the extra tyrosine (des-Tyr mouse relaxin) which had the normal disulfide crosslinking pattern and would be called a revertant by molecular evolutionists. This molecule, when tested side by side with normal mouse relaxin in the receptor-binding assay, proved to be better by about one order of magnitude (Fig. 14.2).[2] The ultimate test was to measure the effect of the derivative on the symphyseal widening in ovariectomized estrogen-primed mice. The result left no doubt that des-tyrosine mouse relaxin was more active than native mouse relaxin and both of them were less than half as active in the mouse as human or rat relaxin (Fig. 14.3). Once this result had been confirmed by several experiments, the message began to sink in; we had improved on nature! This is not like the reports that a hormone from species A is more active in species B than the endogenous one (salmon calcitonin in man for example), but here the endogenous one was remodeled to improve performance by about 50% in the same animal.

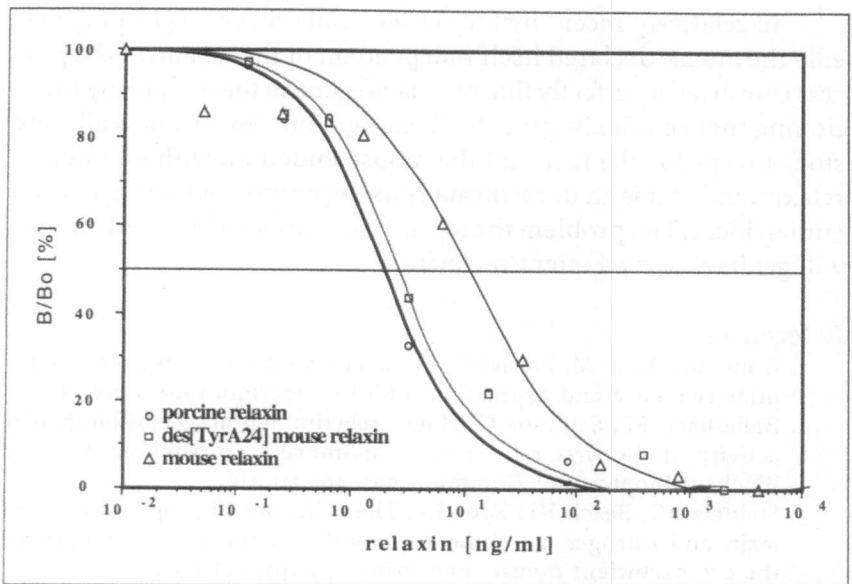

Fig. 14.2. Receptor-binding assay on crude membrane preparations of mouse brain using [125]I-3.5-diiododesamino tyrosyl(A1)porcine relaxin as tracer. Reprinted with permission from Büllesbach EE, Schwabe C. Biochem Biophys Res Commun 1993; 196:311-319.

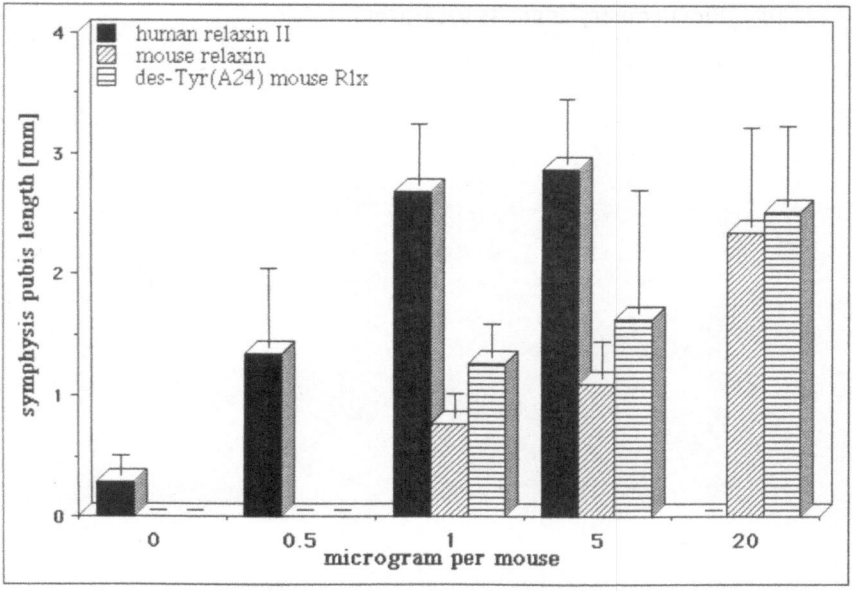

Fig. 14.3. Mouse symphysis pubis assay of mouse relaxin, (des-Tyr)mouse relaxin and human relaxin. The bars indicate standard errors.

In relatively recent history (a few million years ago) purportedly the mouse declared itself independent of the rat and in the process converted a perfectly fine rat relaxin gene to the marginally functioning mouse relaxin gene. Well enough, but this is not really the story except for the fact that the mouse ended up with an inferior relaxin, and that is an unresolvable puzzle for those adhering to Darwinian ideas. The problem though is fascinating and interesting and will get its deserved attention later.

References

1. Evans BA, John M, Fowler KJ et al. The mouse relaxin gene: nucleotide sequence and expression. J Mol Endocrinol 1993; 10:15-23.
2. Büllesbach EE, Schwabe C. Mouse relaxin: Synthesis and biological activity of the first relaxin with an unusual crosslinking pattern. Biochem Biophys Res Commun 1993; 196:311-319.
3. Steinetz BG, Beach VL, Kroc RL. The influence of progesterone, relaxin and estrogen on some structural and functional changes in the pre-parturient mouse. Endocrinology 1957; 61:271-280.
4. O'Byrne EM, Steinetz BG. Radioimmunoassay (RIA) of relaxin in sera of various species using an antiserum to porcine relaxin. Proc Soc Exp Biol Med 1976; 152:272-276.
5. Büllesbach EE, Steinetz BG, Schwabe C. Chemical synthesis of a Zwitterhormon, insulaxin, and of a relaxin-like bombyxin derivative. Biochemistry 1996; 35:9754-9760.

CHAPTER 15

Retro-D-Relaxin

When biochemists use abbreviations such as this title, the meaning is usually discoverable by contemplation (as opposed to the meaning of abbreviations in molecular biology). 'Retro' means the reverse and it is occasionally used exchangeably with 'inverto', and both refer to the direction into which the amino acid backbone is pointing. Peptides are always written with the amino group on the left side and the carboxyl group on the right side, and each single amino acid is incorporated into this peptide in the same direction. The retro- or inverto peptide has each amino acid remaining in the same spot relative to the others, but flipped, so that the carboxyl group now points to the left. Thus, the inverto of Leu→Ala→Gly→Ile would be Leu←Ala←Gly←Ile where the arrows point from the amino terminus to the carboxyl terminus. Using the same example and the conventional way of writing the peptide sequence from the N to the C terminus the retro-form of the original peptide Leu-Ala-Gly-Ile would be Ile-Gly-Ala-Leu. Figure 15.1 explains the concept. The reader will notice that 'retro' does not refer to a natural property of a peptide but rather to one sequence relative to another, i.e., there is no retro peptide without the reference peptide to which it is 'retro'.

The prefix D in this case refers to the non-natural absolute configuration of an amino acid. The configuration of amino acids found in natural proteins is L and biological systems have a very limited number of enzymes that can deal with D-amino acids. For example, no mammalian proteolytic enzyme is known that will attack a peptide in the D-configuration. The objective of this approach is to utilize the biological stability of the D-configuration yet to put all the side chains of the various amino acids into the same three-dimensional space that they occupy in the normal L-form of relaxin. The retro-arrangement then is supposed to compensate for the fact that

Relaxin and the Fine Structure of Proteins, by Christian Schwabe and Erika E. Büllesbach. © 1998 Springer-Verlag and R.G. Landes Company.

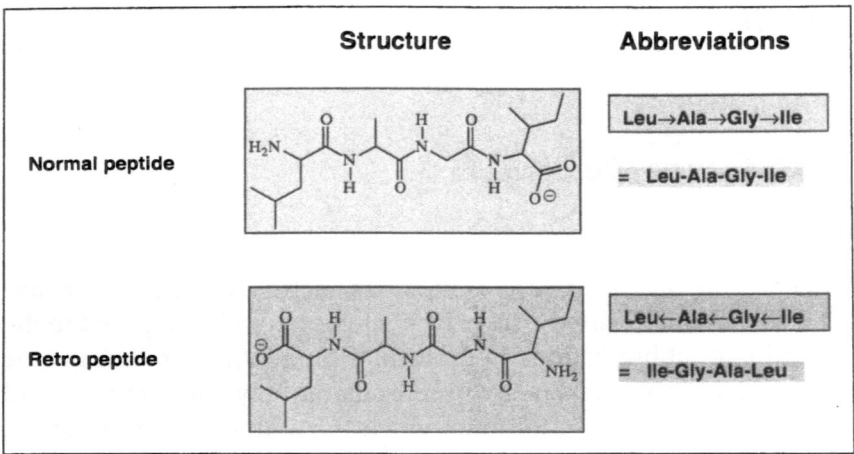

Fig. 15.1. The structure and conventional abbreviations of peptides and retro-peptides.

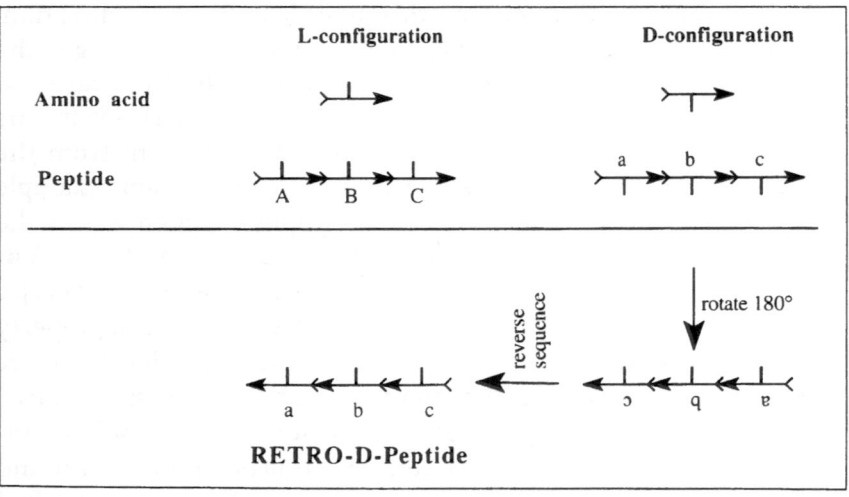

Fig. 15.2. The development of retro-D-peptide and the structural similarity between retro-D and the normal peptide.

the optically active carbon is arranged in space opposite to the L-form. A schematic is given in Figure 15.2.

A hormone made of D-amino acids should be very resistant to enzymatic digestion which in turn would increase its half-life in the circulation. Consequently, lower doses might be given or the administration might be less frequent which would be a distinct advantage to the recipients of hormone therapy. But how should one get the hormone receptor made of L-amino acids to react with D-amino acids in the hormone? As mentioned above the retro-orientation of

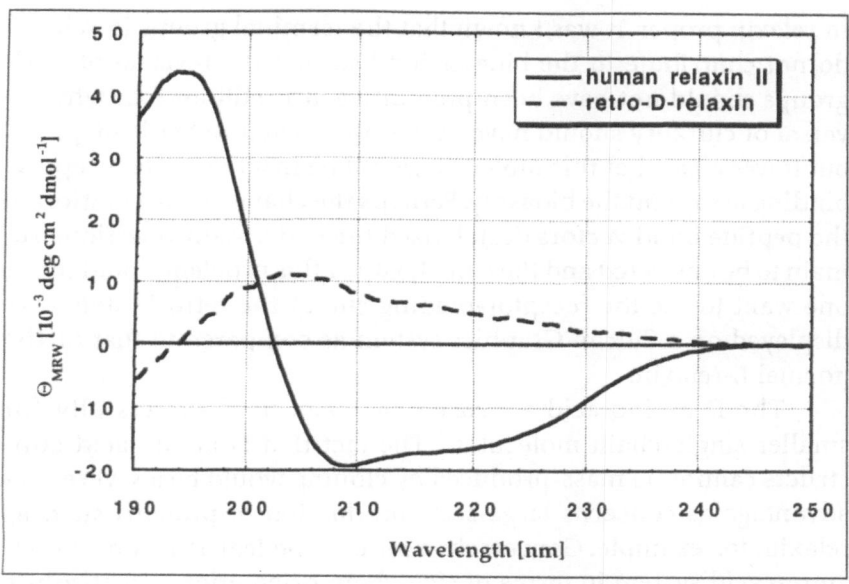

Fig. 15.3. Circular dichroism spectroscopy of human relaxin II (solid line) and its retro-D-form (dashed line).

the amino acids will cause the D-side chain to project into the same space that the L-side chain normally occupies. At least that was the background.

Meanwhile, most amino acid derivatives of the D-configuration had become available commercially in the tBoc and Fmoc form so that the idea could be tested experimentally without having to specifically prepare the protected D-amino acids. The synthesis worked quite well and the site-directed sequential disulfide bond formation produced the properly crosslinked molecule. The CD spectrum showed a clear reversal of the signal as it should be but not the mirror image of L-relaxin that one would expect (Fig. 15.3). Instead the signal is closer to that of a random coil which was a surprise as well as an indication that something had been overlooked in the D-retro transformation; nature is inviting to another humiliating, but likely profitable, lecture on protein structure.

The orientation of the side chains in space in a retro-D peptide mimics the L- peptide but the polarity of the peptide bonds is reversed. All termini are therefore different from those of the natural hormone. The functional group exposed by Asp (B1) for example, is a carboxyl moiety in the retro-relaxin as opposed to an amino group in relaxin proper. It was known that the terminal groups in relaxin do not contribute to the bioactivity[1-3] so that the reversal of end-groups should not have been prohibitive. It is still unclear why reversal of chirality should have caused the structure to disintegrate, but it was clear that this molecule would be inactive in its receptor-binding assay and the bioassay. Perhaps the change in orientation of the peptide bond vectors destabilized the helix. Many questions remain to be answered and the complexity of this problem would make one want to see the receptor-binding site of the retro-D structure displayed on a Silicon Graphics system as compared to that of the normal L-relaxin.

The D-amino acid approach has been used successfully for smaller single-chain molecules.[4] The fact that D-amino acid constructs cannot be mass-produced by cloning would be a severe disadvantage as concerns large-scale production of proteins such as relaxin, for example. Conversely, a lot is to be learned from the D-amino acid system in terms of side-chain orientation and absolute catalytic requirements. Should any of these compounds be extremely useful, methods will be found to produce them *en mass;* that is the way science and technology work together.

The science is not finished; the retro-problem is perhaps 5 strokes from the putting green and needs an investment in patience and understanding to get there.

References

1. Niall HD, John M, James R et al. Structural studies on porcine relaxins and their biosynthetic precursors. In: Brandenburg D, Wollmer A, eds. Insulin: Chemistry, Structure and Function of Insulin and Related Hormones. Berlin: Walter de Gruyter, 1980:719-726.
2. Büllesbach EE, Schwabe C. Naturally occurring porcine relaxins and large-scale preparation of the B29 hormone. Biochemistry 1985; 24:7717-7722.
3. Büllesbach EE, Schwabe C. Preparation and properties of porcine relaxin derivatives shortened at the amino terminus of the A chain. Biochemistry 1986; 25:5998-6004.
4. McDonnell JM, Beavil AJ, Mackay GA et al. Structure based design and characterization of peptides that inhibit IgE binding to its high-affinity receptor. Nature Structural Biology 1996; 3:419-426.

CHAPTER 16

The Relaxin-Like Factor

Testicular Leydig cells contain the message for a protein of the insulin/relaxin-like family which, in the heat of discovery, was called the Leydig insulin-like peptide (LEY I-L)[1] or INSL3, corresponding to the chromosomal location of the human gene.[2] The authors felt that this was a Leydig cell specific message and predicted an insulin-like character for the gene product.[1] Upon close scrutiny one will notice that the critical position in the A chain-loop, which decides between insulin or relaxin-like conformations, was occupied by glycine (relaxin configuration) instead of isoleucine (insulin configuration). Furthermore, the cDNA sequence of the B-chain also showed more relaxin-like characteristics, in particular the binding-site sequence RXXXR which occurred four residues further toward the C-terminal end of the molecule (Fig. 16.1). This meant a displacement of exactly one helix turn with the arginines still projecting essentially away from the core as they do in relaxin. All this was enough reason to synthesize the molecule according to the cDNA sequence and to study its physicochemical and biological properties. Within a month this new factor was synthesized and measurements of circular dichroism as a function of wavelength showed a conformation very similar to that of porcine relaxin.[3] A tracer was produced by direct iodination of a synthetic precursor as given below.

The synthetic RLF was used at the stage where the tryptophan and methionine were still side chain-protected. The peptide, 10 µg in 5 µl water, was pipetted into a 500 µl Eppendorf vial and chilled on ice. In quick succession 5 µl of phosphate buffer (250 mM, pH 7.4) was added followed by 2 µl of $Na^{125}I$ (1 mCi) and 5 µl of chloramine T (2 mg per ml in phosphate buffer, pH 7.4). The reaction proceeded for one minute on ice and was then quenched with 5 µl of sodium thiosulfate (5 H_2O) (50 mg/ml in phosphate buffer, pH 7.4), and 5 µl

Relaxin and the Fine Structure of Proteins, by Christian Schwabe and Erika E. Büllesbach. © 1998 Springer-Verlag and R.G. Landes Company.

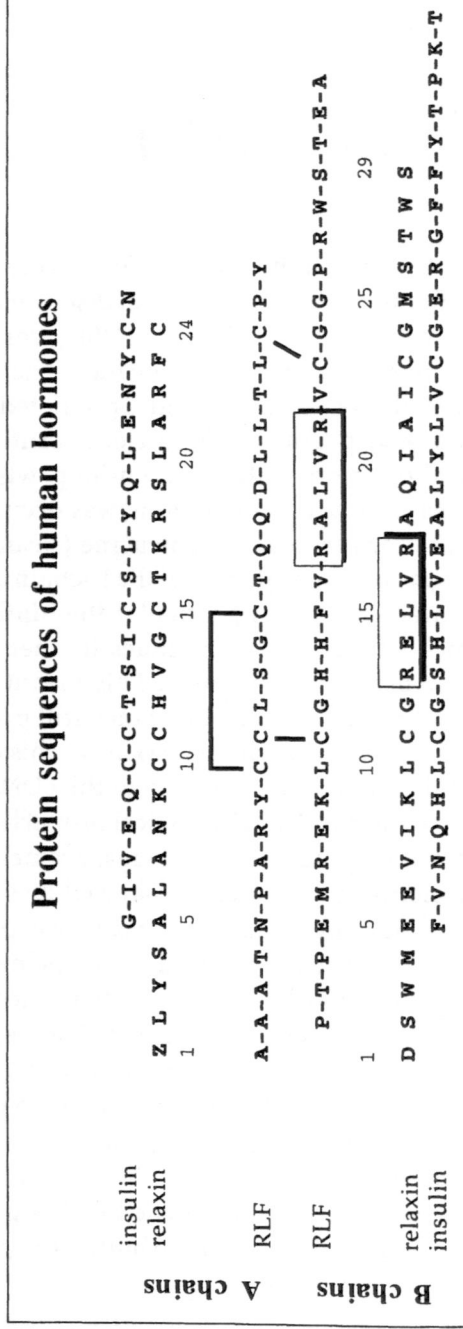

Protein sequences of human hormones

A chains

insulin   G-I-V-E-Q-C-C-T-S-I-C-S-L-Y-Q-L-E-N-Y-C-N

relaxin   Z-L-Y-S-A-L-A-N-K-C-C-H-V-G-C-T-K-R-S-L-A-R-F-C
           1   5   10   15   20   24

RLF   A-A-A-T-N-P-A-R-Y-C-C-L-S-G-C-T-Q-Q-D-L-L-T-L-C-P-Y

B chains

RLF   P-T-P-E-M-R-E-K-L-C-G-H-H-F-V-R-A-L-V-R-V-C-G-G-P-R-W-S-T-E-A
            1   5   10   15   20   25   29

relaxin   D-S-W-M-E-E-V-I-K-L-C-G-R-E-L-V-R-A-Q-I-A-I-C-G-M-S-T-W-S

insulin   F-V-N-Q-H-L-C-G-S-H-L-V-E-A-L-Y-L-V-C-G-E-R-G-F-F-Y-T-P-K-T

Fig. 16.1. Comparison of the primary structures of human relaxin-like factor (RLF) with human relaxin II and human insulin. The proposed active site of relaxin is boxed in and compared with RLF in which a similar motive is offset by four residues toward the C terminus of the B chain.

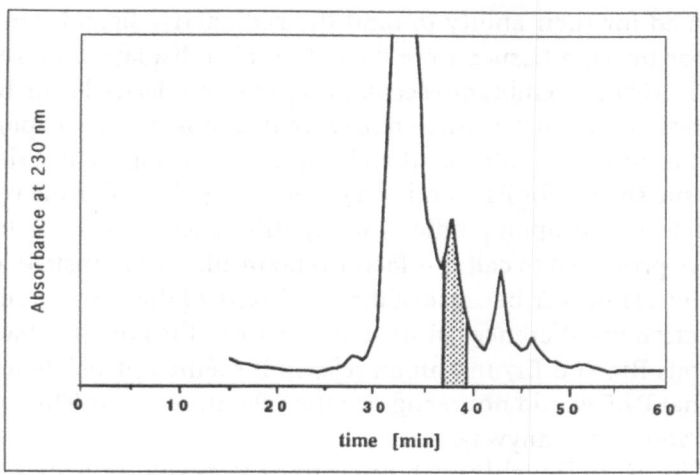

Fig. 16.2. Radioactive iodination of RLF. HPLC profile of the reaction mixture on an Aquapore 300 (2.1 x 30 mm) column. The solvent system consisted of 0.1% trifluoroacetic acid in water (solvent A) and 0.1% trifluoroacetic acid in 80% acetonitrile (solvent B). The column was equilibrated at 23% B and the protein eluted with linear gradient from 23% B to 34% B in 60 min. Thereafter the concentration of B was increased to 100% within a minute, kept at 100% for four minutes before changing to initial conditions within 1 min. The main peak eluting at 30 minutes is the sulfoxide of RLF. The shaded area was collected and used as RLF tracer. This tracer contains one ^{125}I in the A chain either in position A9 or A26 and a sulfoxide in methionine B6. Reprinted with permission from: Büllesbach EE, Schwabe C. J Biol Chem 1995; 370:16011-16015.

of NaI (20 mg/ml in phosphate buffer, pH 7.4). The side chain-protecting group of tryptophan was removed by adding 5 µl of piperidine. After two minutes at room temperature the reaction was quenched with 5 µl of glacial acidic acid, the mixture diluted with 10 µl of water and loaded onto an Aquapore 300 column for separation (Fig. 16.2). The protein was detected by UV absorbance and manually collected in tubes containing 100 µl of 1% bovine serum albumin in water. RLF tracers still contain methionine sulfoxide.[3]

Human RLF has two tyrosine residues in the A-chain (A9 and A26) which can be labeled. The relatively hydrophilic unlabeled sulfoxide of RLF eluted first from the column (Fig. 16.2). Two subsequent fractions of similar UV absorbance and radioactivity were collected and used side by side during the initial binding studies. The first eluting radioactive isomer showed less nonspecific binding and was therefore the tracer of choice. With this tracer several tissues such as brain, uterus, liver, kidney and skeletal muscle were examined for their ability to bind the radioactive ligand. The RLF receptor-bearing tissues were those that also displayed the relaxin receptor. Crude membrane receptor preparations derived from brains and uteri of estrogen-primed mice revealed more receptors per milligram of protein in uterine than in brain tissue. Figure 16.3 depicts the tissue survey for RLF and relaxin which again underline relaxin affinities. Based upon studies with synthetic RLF and RLF-derivatives we proposed to call this factor relaxin-like (RLF) instead of insulin-like. However, because of the sensitivity of the relaxin receptor interaction to differences in size and nature of the positive charge in positions B13 and B17 in human relaxin it seemed plausible to concede that RLF would not recognize the relaxin receptor. The experiments were done anyway.

The unbelievable, yet unequivocal result revealed weak crossreactivity of RLF to the relaxin receptor and it seemed justified to suspect that the RLF arginines were responsible. There is no compelling explanation for this observation. Nature seems to be intolerant of the slightest attempt by protein chemists to escape empiricism. In light of all the experimental evidence concerning the sensitivity of receptor-binding to the presence of unmodified arginines (chapter 6), the reader may agree that the prediction was reasonable and that crossreactivity, however slight, came as a surprise. It is possible that the RLF shifts in toto four residues along its B-chain helix to match the binding-site or that it turns 180° to bind in the retro-position. This would not be such a bad proposition if receptor-binding were not influenced by other features in the B-chain. The importance of other residues can be inferred from the fact that all relaxins have the two arginines yet their receptor-binding propensity differs by several orders of magnitude.

With its slight crossreactivity to the relaxin receptor could unmodified RLF be a relaxin inhibitor? A series of bioassays were set up to answer this question. The first experiment was designed to establish a baseline with a control sample of 1 μg of human relaxin, or 5 μg and 20 μg of RLF alone. In the last group of animals, 1 μg of human relaxin plus 5 μg of RLF were given together. In the next experiment relaxin was increased whereas RLF was kept constant at 5 μg per animal. The final experiment was performed with a constant human relaxin level of 5 μg and an increasing RLF dose for succeeding groups. These data show that RLF by itself has no effect on the symphysis pubis in estrogen-primed mice. The data suggest

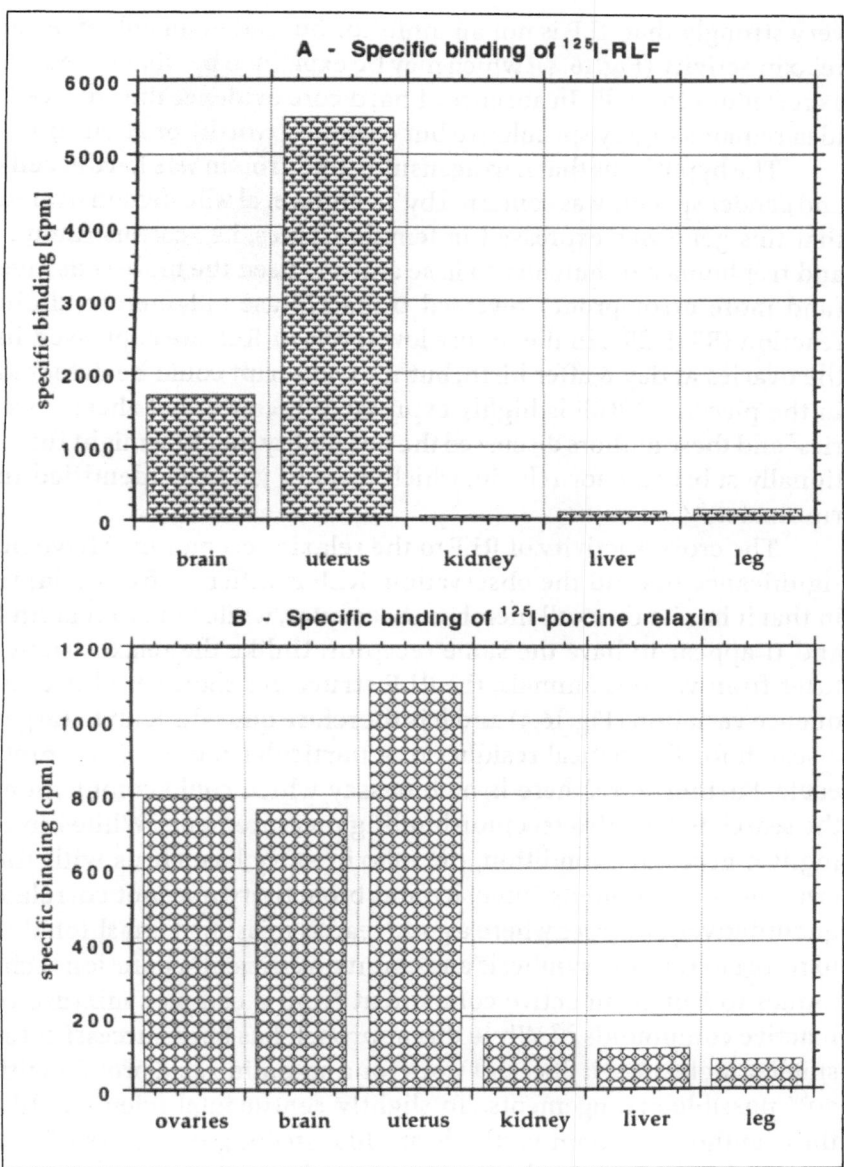

Fig. 16.3. Tissue specificity of RLF (panel A) versus relaxin (panel B) (relaxin data were adopted from reference 13 and RLF data from ref. 3). The tissue distribution of receptors in female estrogen-primed mice as measured on crude membranes in a receptor-binding assay. Reprinted with permission from: Büllesbach EE, Schwabe C. J Biol Chem 1995; 370:16011-16015 and Yang S, Rembiesa B, Büllesbach EE et al. Demonstration of ligand binding in symphyseal tissues and uterine membrane. Endocrinology 1992; 130:179-185. © The Endocrine Society.

very strongly that RLF is not an inhibitor but rather an enhancer of relaxin activity (Fig. 16.4)[3] which may be explained by what is known as receptor crosstalk. In absence of hard core evidence the crosstalk idea remains highly speculative but certainly worthy of attention.

The hypothesis that Leydig insulin-like protein was Leydig cell- and gender specific was contested by Tashima et al who demonstrated that this gene was expressed in female tissues, i.e., corpus luteum and trophoblast of humans.[4] These authors used the more sensitive (and more error-prone) reversed transcriptase polymerase chain reaction (RT-PCR). In the mouse low levels of RLF are expressed in the ovaries at day 6 after birth, but no transcript could be detected in the placenta.[5] RLF is highly expressed in cow[6,7] and sheep ovaries[7] and these authors discussed the possibility that RLF might functionally substitute for relaxin, which has not yet been identified in ruminates.[6,7]

The crossreactivity of RLF to the relaxin receptor may have no significance beyond the observation. RLF is different from relaxin in that it has its own cell membrane receptor, while human relaxin I and II appear to have the same receptor. Unlike the relaxin structures from various animals, the RLF structures show very little sequence variation (Fig. 16.5) and it is therefore quite difficult to target a search for the critical residues to a particular region of the molecule. Furthermore, there is no bioassay which could complement the search for in vitro receptor-binding measurements. While binding is a necessary condition, the relaxin story has left us with the awareness that binding intensity and bioactivity may not correlate quantitatively. In cases where a good lead is missing it is fashionable to resort to random synthesis and count on efficient separation techniques to isolate an active component from a complex mixture of inactive compounds.[8,9] While this approach has been successful for smaller molecules, it does not work for a 50-amino acid protein with 20^{50} possible arrangements. To slightly sentimental scientists like these authors, random methods are like striking the colors of science and relinquishing the field to technology. Giving up on the attempts to understand and target one's action for the "try all on the shelf" approach seems like returning to the prescience era.

A compromise method, i.e., a negotiated surrender to ignorance, is called alanine scanning. One residue after another in RLF, for example, would be replaced by alanine under this protocol and the response of a molecule to the substitution is tested in an assay system. Wherever alanine substitution leads to a change a residue is deemed important for activity. The reader who has gone through the chapters of relaxin structure-function studies will remain skeptical about the interpretation of an alanine scan.

How then would we find the active site with little hint from structural differences of RLF from various species? It was clear from the onset that many RLF derivatives would have to be prepared before one could hope to understand the RLF molecule and its surface as we understand relaxin. Of the five different RLF structures known the largest difference (26%) was observed between mouse and human RLF (Fig. 16.5).[5,10,11]

The search for the active site is on and human RLF has been selected as reference. When synthetic mouse RLF was compared with human RLF it showed clearly that both molecules were equipotent. Thus, it was possible to eliminate few amino acid side chains from the pool of possible active-site residues. Besides that, there was no terribly obvious target region. The first attempts were therefore guided by practical considerations, in particular technical aspects, that would simplify the synthesis of derivatives to be used for receptor identification. The curse of peptide chemistry is tryptophan which requires a great deal of care and often extra steps to avoid oxidation of the indole ring. For that reason it was decided to test tryptophan as regards receptor-binding activity of RLF and to see if it could be exchanged for a more user-friendly residue. The flexibility of solid phase synthesis made possible simultaneous substitution of tryptophan in position B28 with four different amino acids. To achieve this goal the RLF B-chain was synthesized up to position 29 and the resin was then divided into four different portions. Each portion was used to introduce a different amino acid into position B28, i.e., Phe, His, Leu, and Ala. Then the four portions were pooled and the synthesis continued according to the native sequence. At the end four RLF B chains had been obtained with tryptophan in position B28, replaced by phenylalanine, leucine, histidine and alanine, respectively. The four chains were separated by preparative high performance liquid chromatography (HPLC), each individual chain combined with the A chain, yielding four RLF derivatives which were then compared to native RLF as regards structure and receptor-binding.

Biological research is characterized by surprises, good or bad. While the good ones are preferred all of them build up to whatever is fascinating in science and this time was no different. Tryptophan

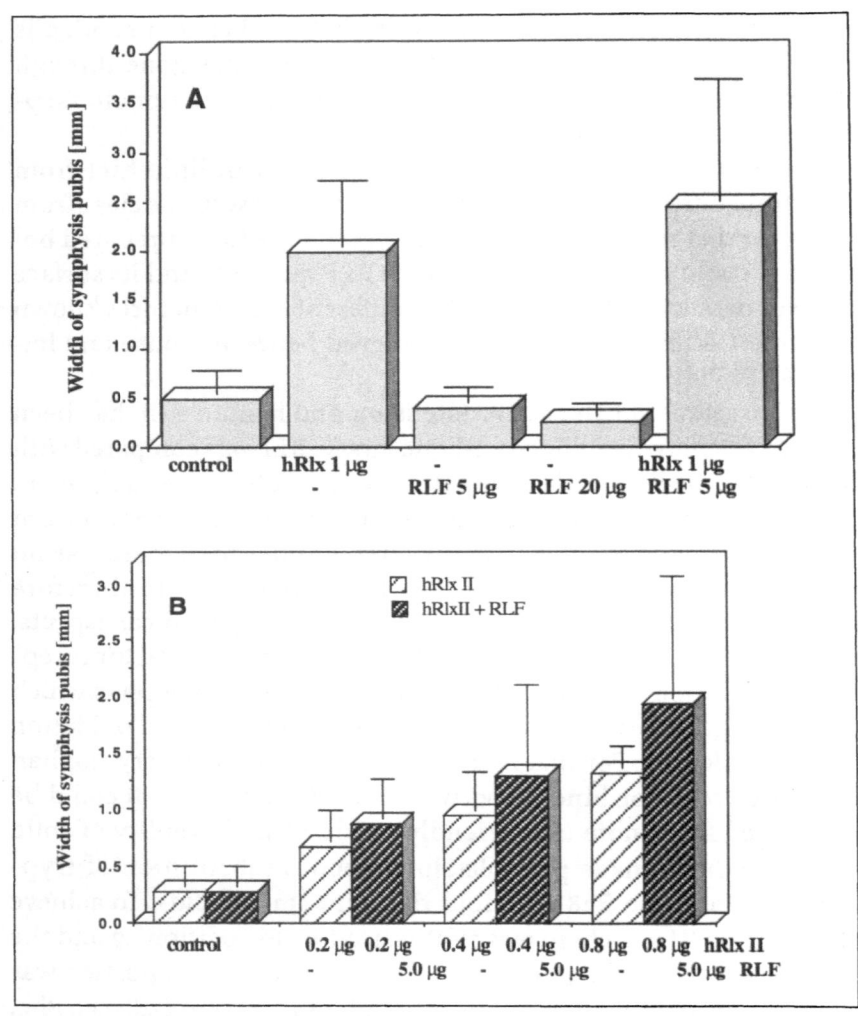

Fig. 16.4. The effect of RLF in the mouse symphysis pubis assay. (A, above) Assay of the individual hormones relaxin and RLF and a mixture of both. RLF alone has no effect on the symphysis pubis while the combination of RLF and relaxin seemed to increase the relaxin effect at optimal dose. (B, above) Dose response of relaxin in the presence and absence of RLF (5 μg/mouse). The response in the presence of RLF is higher than the response with relaxin alone. (C, opposite) Dose response of RLF in the presence of a uniform level of human relaxin. With increasing dose of RLF the response on the symphysis pubis increases. Bars indicate standard deviation. Reprinted with permission from: Büllesbach EE, Schwabe C. J Biol Chem 1995; 370:16011-16015.

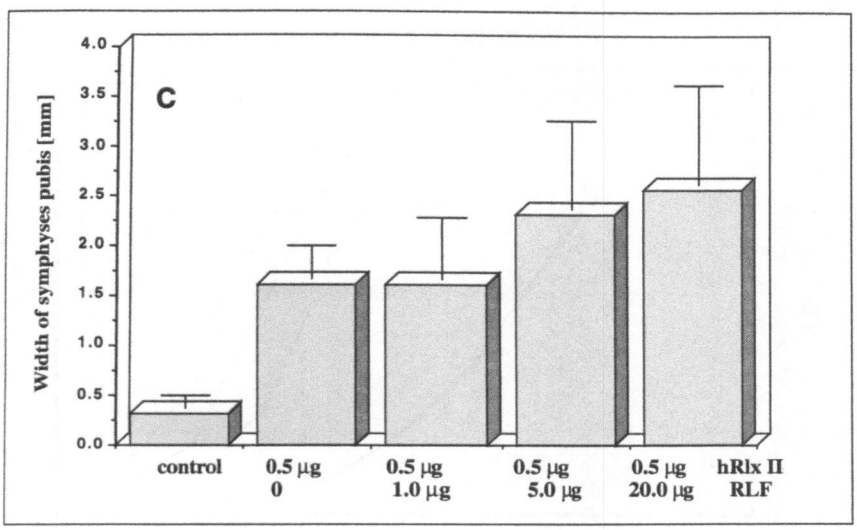

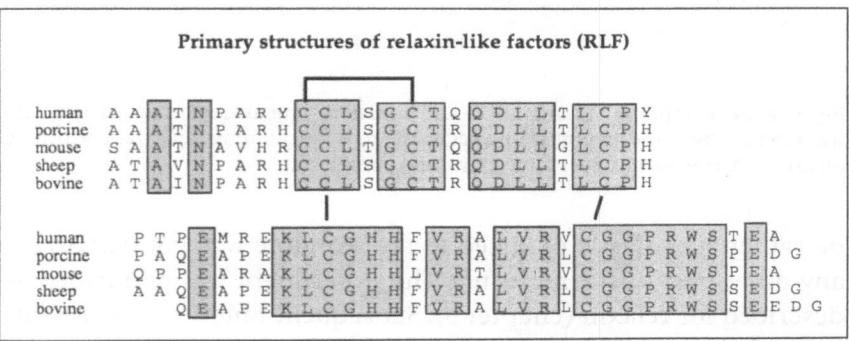

Fig. 16.5. Sequence comparison of relaxin-like factors. All sequences were deduced from cDNA sequences. The shaded area indicates the constant amino acids in all five sequences.

appeared to be very important for function and while that wrecked havoc with the synthesis plans it gave in return a very strong hint that the receptor-binding region of RLF had been stumbled upon. None of the substituted amino acids would come close to native RLF, although the phenylalanine(B28) RLF did a little better then the rest (Fig. 16.6).

Tryptophan is not often considered a mediator of biological activity and that may still be true in this case. Receptor-binding however is distinctly influenced by the sole tryptophan in RLF, leaving one curious whether the specific effect is due to the aromatic character of the indole ring or the size of the side chain. Tryptophan can

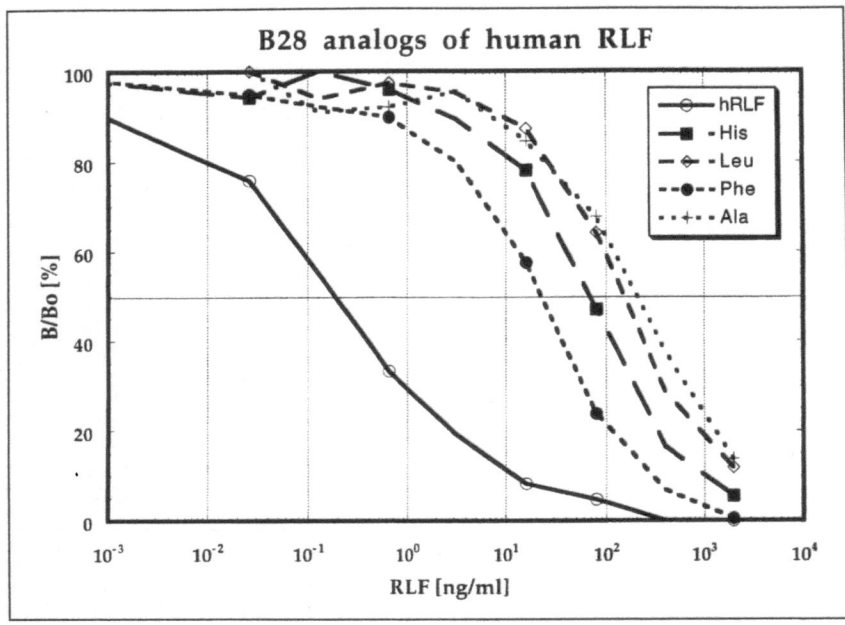

Fig. 16.6. Dose response of human RLF and human relaxin analogs in which Trp(B28) was replaced by four different amino acids. The receptor-binding assay was performed on crude uterine membranes from estrogen-primed mice.

be rather specifically oxidized to the oxindole without destroying any other part of the molecule. The experiment was performed as described for relaxin (chapter 6). Subsequent binding-experiments showed significant reduction in receptor-binding which suggested that the aromatic quality of the indole ring must be maintained in order to maximize receptor binding. Suddenly the monotonous sequences of RLF came to life with suggestions as to its active region. For example, there are significant structural determinants in the immediate vicinity of tryptophan B28. The GGP sequence, for example, dictates a bend just after the last cysteine whereas the rest of the B chain C-terminal region is long enough to coil up as an α-helix. The bend sequence GGP has suddenly become a logical target for the binding region.

The search for a practical advantage may have provided, inadvertently, exceedingly important information concerning this new factor. There will be more on this aspect in the near future.

Another analog of interest was a photo-crosslinkable radioactive RLF to be used for receptor isolation and characterization. Suspecting that the C-terminal end of the B chain is near the receptor-binding site the photo-activatable crosslinker p-benzoyl-phenylalanine (Bpa) was incorporated in position A1 in exchange for alanine. This derivative should retain high affinity for the receptor and provide an RLF with a radioactive label close to the crosslinking moiety. Indeed, receptor-binding assays of this RLF derivative revealed that the first prediction was correct; Bpa(A1) RLF bound the receptor as well as native RLF, indicating that the N-terminal variable region of the A chain is not part of the RLF receptor interaction site and can tolerate a fairly bulky amino acid. The crucial question was whether or not the Bpa side chain is in contact with the receptor. To answer this question radioactive Bpa(A1)-RLF needs to be prepared, the hormone receptor complex formed, photocrosslinked, and analyzed. Again the edge of the RLF world has been reached.

Relaxin and RLF are representatives of two different generations of biomolecular research; relaxin developed historically from an observed biological effect to a molecular structure with eventual pharmaceutical application while the relaxin-like factor was at first a gene without known biological function that induced curiosity. Only after the molecule became available through chemical synthesis could one begin to search for receptors and thus for a possible sphere of influence. Augmentation of some of the relaxin effect had been observed and the likely role in sperm development was a popular speculation until recently when the studies of Adham and colleagues provided strong evidence for a role in testicular physiology during the early stages of spermatogenesis and germ-cell maturation.[1,5] RLF gene deletion experiments produced premeiotic arrest and germ cell death.[12] Furthermore, the possibility of RLF action in the ovaries should not be discarded as yet. RLF is expressed in mouse ovaries 6 days after birth and at various stages of the estrous cycle and during pregnancy, suggesting that RLF may also be involved in follicle development.[5] Whether or not RLF replaces the missing relaxin in ruminants remains speculative[6,7] but it cannot be dismissed out of hand since potential target organs for RLF (in the mouse) are also receptor-bearing tissues for relaxin.[3] In fact our latest research results suggest rather strongly that RLF is a circulating hormone in males and females.

It may be a personal bias at this time, but these authors do predict with conviction that RLF (under any name) will become an important factor in modern physiology and medicine.

References

1. Adham IM, Burkhardt E, Benahmed M et al. Cloning of a cDNA for a novel insulin-like peptide of the testicular Leydig cells. J Biol Chem 1993; 268:26668-26672.
2. Burkhardt E, Adham IM, Brosig B et al. Structural organization of the porcine and human genes coding for a Leydig cell-specific insulin-like peptide (LEY I-L) and chromosomal localization of the human gene (INSL3). Genomics 1994; 20:13-19.
3. Büllesbach EE, Schwabe C. A novel Leydig cell cDNA-derived protein is a relaxin-like factor (RLF). J Biol Chem 1995; 370:16011-16015.
4. Tashima LS, Hieber AD, Greenwood FC et al. The human Leydig insulin-like (hLey i-l) gene is expressed in the corpus luteum and trophoblast. J Clin Endocrinol Metab 1995; 80:707-710.
5. Zimmermann S, Schottler P, Engel W et al. Mouse Leydig insulin-like (Ley I-L) gene—Structure and expression during testis and ovary development. Mol Reprod Develop 1997; 47:30-38.
6. Bathgate R, Balvers M, Hunt N et al. Relaxin-like factor gene is highly expressed in the bovine ovary of the cycle and pregnancy—Sequence and messenger ribonucleic acid analysis. Biol Reprod 1996; 55:1452-1457.
7. Roche PJ, Butkus A, Wintour EM et al. Structure and expression of Leydig insulin-like peptide mRNA in the sheep. Mol Cell Endocrinology 1996; 121:171-177.
8. Plunkett MJ, Ellman JA. Combinatorial Chemistry and New Drugs. Scientific American 1997; 276:68-73.
9. Hogan JC. Combinatorial Chemistry In Drug Discovery. Nature Biotechnology 1997; 15:328-330.
10. Burkhardt E, Adham IM, Hobohm U et al. A human cDNA coding for the Leydig insulin-like peptide (LEY I-L). Hum Genet 1994; 94:91-94.
11. Pusch W, Balvers M, Ivell R. Molecular cloning and expression of the relaxin-like factor from the mouse testis. Endocrinology 1996; 137:3009-3013.
12. Adham IM, Zimmermann S, Engel W. Leydig insulin-like hormone-deficient mice show male premeiotic arrest and germ cell death. Reprod Domestic Animals 1997; 32:73.
13. Yang S, Rembiesa B, Büllesbach EE et al. Relaxin receptors in mice: Demonstration of ligand binding in symphyseal tissues and uterine membrane. Endocrinology 1992; 130:179-185.

CHAPTER 17

The Relaxin Receptor

The relaxin receptor has been elusive for an unusually long time
and still is not terribly forthcoming with detailed information.
While the interaction with its natural ligand is quite strong the high
nonspecific binding, which is a property of relaxin, has made the
usual radioligand-mediated detection methods quite difficult. In
contrast, insulin has low nonspecific binding, but shows specific binding up to 80% of the total radioactivity bound. The relaxin system
gives 50% specific binding at best and usually shows around 30%,
and that only after major efforts in tracer design, preparation and
purification.

The first evidence that relaxin targets reproductive organs was
obtained by Cheah and Sherwood.[1] Radioactive porcine relaxin tracer
(5 microcuries = 10^7 dpm) was injected into the tail veins of estrogen-primed rats. Control animals received a 1000-fold excess of unlabeled porcine relaxin over tracer. Animals were sacrificed at certain intervals for blood and tissues collection. Tissue weights were
recorded and radioactivity was measured in a gamma-counter. When
counts per 100 µl blood were subtracted from the counts per 100 mg
tissue, several organs showed accumulated radioactivity, i.e., thyroid,
spleen, kidney, liver, lung and uteri, but only the binding to uterine
tissue was reduced significantly in the presence of porcine relaxin.[1]
This observation was in harmony with subsequent in vitro binding-
studies.

The classical target organ of relaxin action is the symphysis
pubis, yet it took very sensitive radio-histochemical methods to demonstrate the presence of specific relaxin receptors. For these studies
HPLC-purified monocomponent ^{125}I-relaxin tracer with a specific
activity of 4.4 Ci/µmol was used in combination with hypersensi-
tive film emulsion.[2]

Relaxin and the Fine Structure of Proteins, by Christian Schwabe
and Erika E. Büllesbach. © 1998 Springer-Verlag and R.G. Landes Company.

Virgin female ICR mice (18-20 g) were purchased from Charles River, Wilmington, MA, for these experiments. The mice were estrogen-primed by an injection of 5 µg depot estrogen in 100 µl sesame oil. Five days later the animals were killed in an atmosphere of CO_2, the tissues were excised and immediately embedded in OCT (Miles, Inc., Elkhart, IN) and maintained at -20°C. About 10 µm sections were cut on a Minitome (International Equipment Co., Needham Heights, MA) and deposited on microscope slides which were pretreated with 0.3% gelatin and 0.01% chromium potassium sulfate for better tissue retention. The OTC-embedded sections were warmed briefly to 37°C. Immediately after drying consecutive tissue sections were covered with 100 µl of either binding buffer alone or binding buffer containing 10 ng of porcine relaxin and preincubated for 30 min. Thereafter the tracer was added (30,000 cpm/section) and all sections were incubated at room temperature in a moist chamber for 30 minutes, then washed in buffer followed by water, and partially dehydrated by dipping in 30% and 60% ethanol, before air drying. Coating with NTB2 nuclear track emulsion was achieved by dipping the slides into the emulsion in total darkness. After 24 hour-exposure the slides were developed in Kodak developer D-19, fixed in Kodak fixer at 15°C and washed in distilled water at the same temperature. Tissue outlines were produced by subsequent Giemsa stain.[2]

The relaxin receptors are located on the interpubic ligament of the symphysis pubis (Fig. 17.1A,B). In addition uteri, ovaries and leg muscles were dissected and treated simultaneously. Distinct differences between tissues treated with radioactive relaxin alone and radioactive relaxin in the presence of a 340-fold excess of porcine relaxin were observed for uteri and ovaries; no difference was detected for leg muscles. In uteri the relaxin receptor is located in the epimetrium where it appears in clusters (Fig. 17.1C,D), and in ovaries a strong signal is seen in the peripheral layers (Fig. 17.1E,F).[2]

Relaxin receptors should be located on the cell surface and, accordingly, crude membranes were prepared from potential target tissues and purported non-target tissues as controls. For target tissue, ovaries and uteri of estrogen-primed mice were used. Crude membranes were prepared according to standard protocols[3] and specific binding was determined by exposing the crude membranes to tracer (100 pM) in the presence or absence of cold porcine relaxin (0.3 µM). After a certain time unbound tracer was separated by centrifugation and aspiration, and the pellet was transferred to a

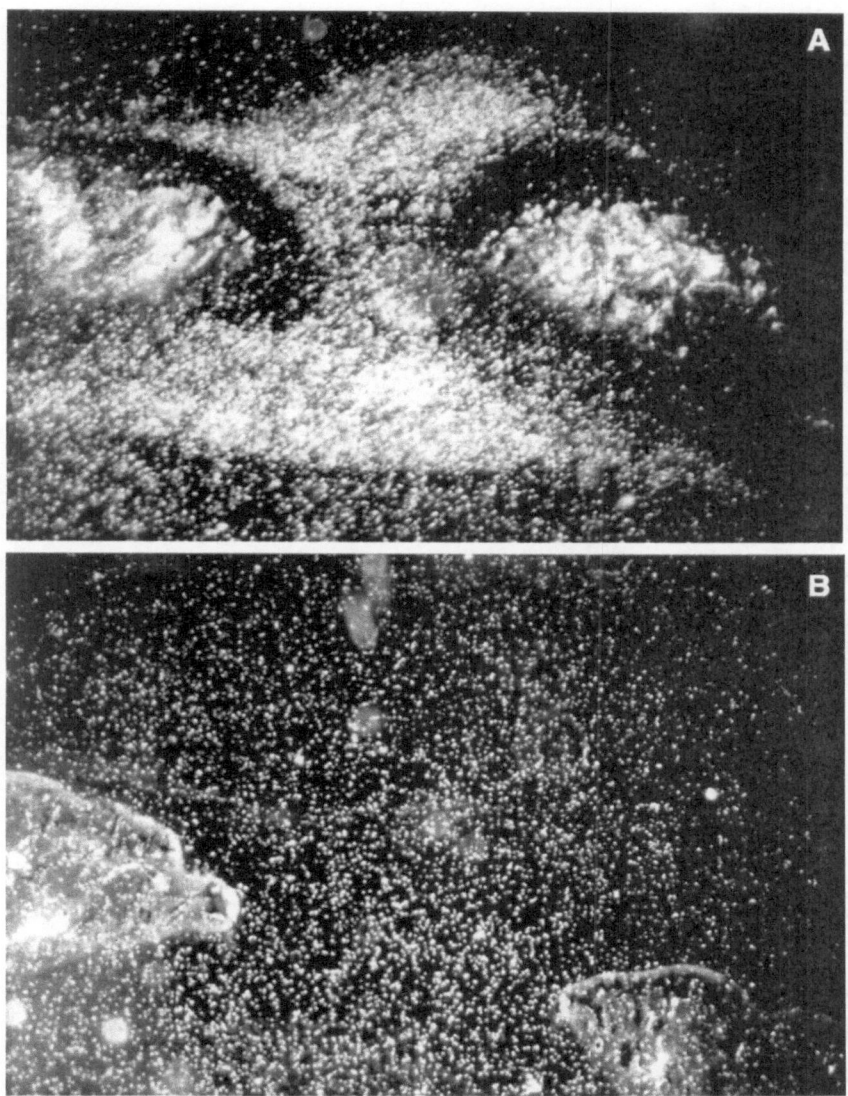

Fig. 17.1. Radio-histochemical identification of relaxin receptors in mouse tissues. Tissue slides were treated with binding buffer in the presence of 30,000 cpm of porcine relaxin tracer (experiment) or pretreated with 10 ng of porcine relaxin followed by incubation with 30,000 cpm of tracer (control). (A) symphysis pubis (experiment); (B) symphysis pubis control; (C) uterus experiment; (D) uterus control; (E) ovary experiment, and (F) ovary control. Reprinted with permission from Yang S, Rembiesa, B, Büllesbach EE et al. Endocrinology 1992;130:179-185. (Panels C-F on following pages.)

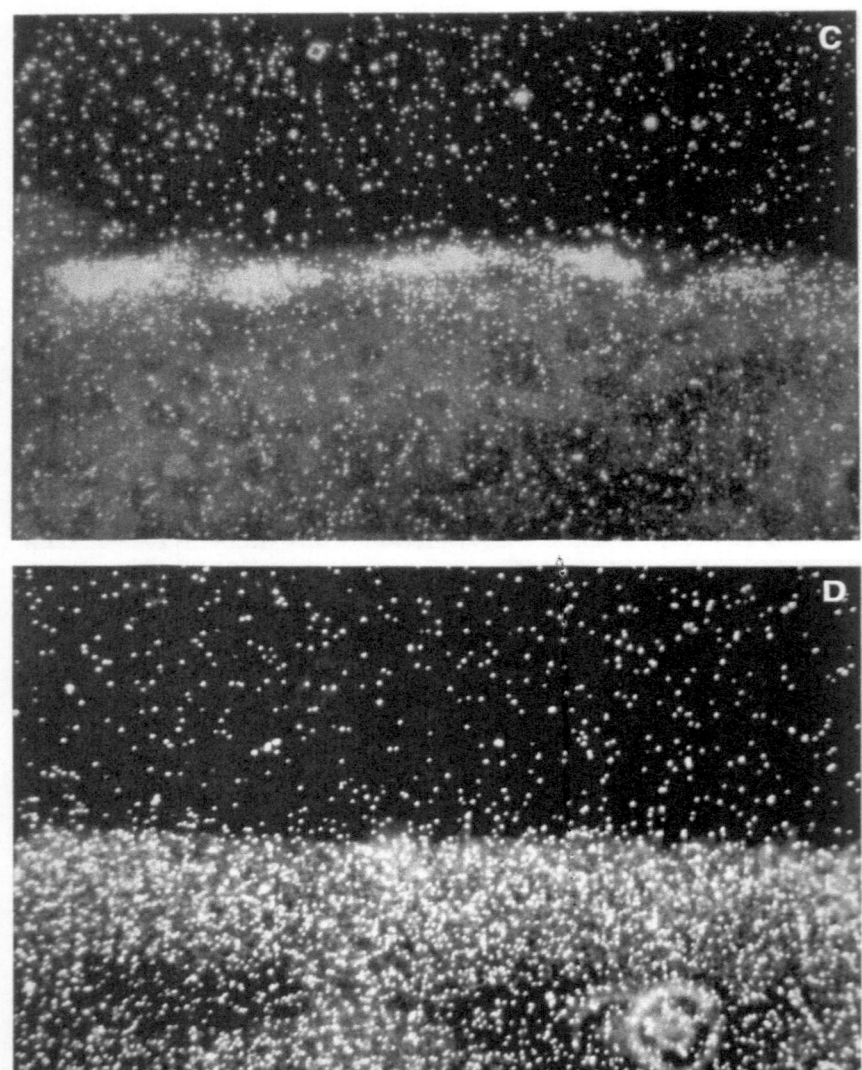

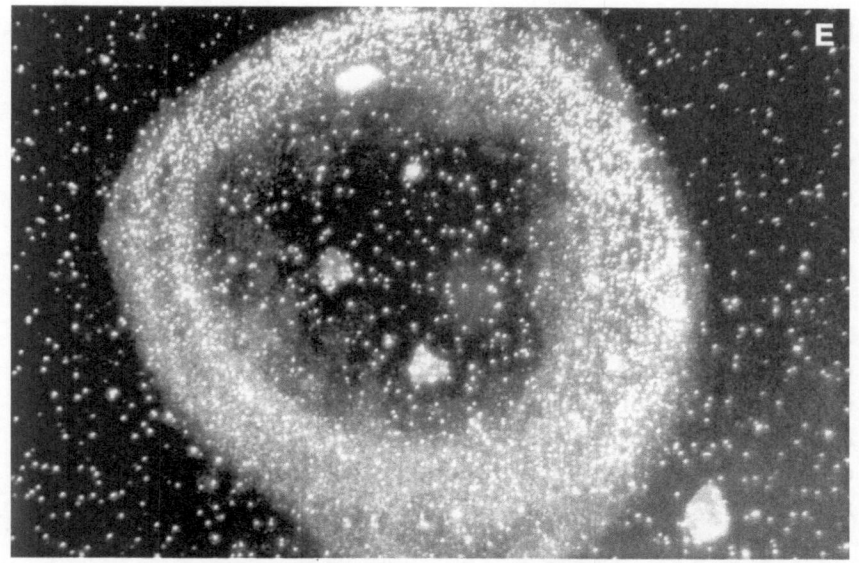

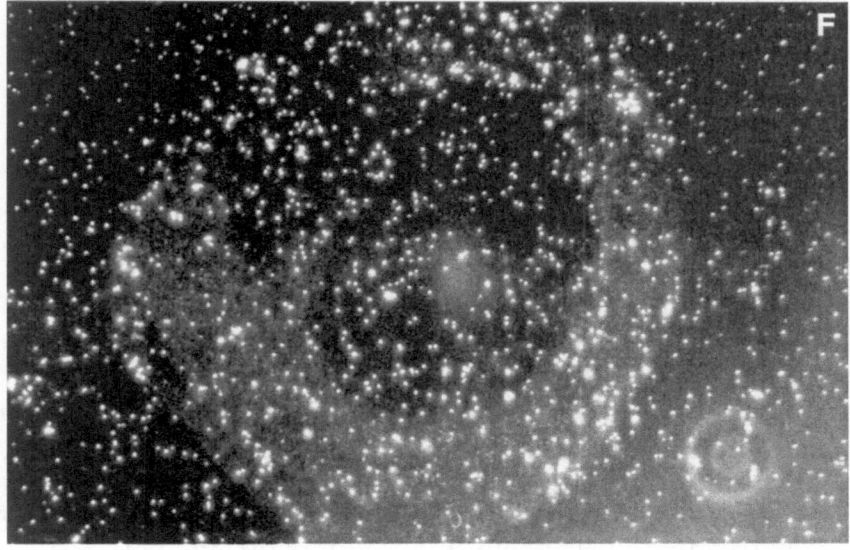

gamma-counter. To optimize binding-conditions only total binding and nonspecific binding were determined in quintuplicates for each data point on uterine tissues and leg muscles, and promising experiments were repeated independently at least three times. The goal was to increase total binding and to decrease nonspecific binding. Critical to initial positive results was the tracer (chapter 7), the presence of serine proteinase inhibitors in all buffers, and the addition of bivalent ions to the binding buffer. Technical details, like avoiding vortexing and the use of chilled wash buffers, were important as well. All improvements combined boosted specific binding to 50% of the total ligand bound, a significant improvement when compared to the first reports on relaxin receptors which showed five percent specific binding.[4] Using ^{32}P labeled relaxin (see below) Garibay-Tupas et al were able to demonstrate good specific binding in human fetal membranes.[5]

Receptor-binding assays on crude membrane preparations of mouse tissue: Two whole mouse brains (or uteri of 5 estrogen-primed mice) were extirpated and dropped into 15 ml of chilled buffer (25 mM HEPES, 0.14 M NaCl, 5.7 mM KCl, 0.2 mM PMSF, and 80 mg/l soybean trypsin inhibitor, pH 7.5), supplemented with sucrose (0.25 M), and homogenized three times for 10 seconds with a Polytron homogenizer (Brinkmann, Westbury, NY) at setting 5. The homogenate was centrifuged at 700 x g for 10 min at 4°C, the pellet was again homogenized in 15 ml of sucrose-containing buffer and centrifuged under the same conditions. The pellet was discarded and the supernatants combined and centrifuged for 1 h at 20,000 x g. The resulting pellet was suspended in buffer without sucrose addition and centrifuged again for 1 h at 20,000 x g.

For binding-assays crude membranes of two mouse brains (or 5 uteri) were suspended in 1 ml of binding buffer (25 mM HEPES, 0.14 M NaCl, 5.7 mM KCl, 0.2 mM PMSF, 1% BSA, 2.8 mM glucose, 1.6 mM $CaCl_2$, 0.025 mM $MgCl_2$, and 1.5 mM $MnCl_2$). Aliquots of 40 µl of crude membranes were added to 40 µl of ^{125}I-diiododesaminotyrosyl(A1)-porcine relaxin (100,000 cpm) and 20 µl of buffer (total binding) or 20 µl of the corresponding relaxin analog (dose response) in 1.5 ml Eppendorf vials. The assay was incubated for 1 h at room temperature, then diluted with 1 ml of ice-cold wash buffer (25 mM HEPES, 0.14 M NaCl, 5.7 mM KCl, 1% BSA, 0.01% NaN_3), and centrifuged for 10 min at 14,000 rpm at room temperature. Thereafter the supernatant was discarded and the tip of the vial cut off and counted in a gamma counter (Minigamma, LKB, Sweden). Remaining radio-activity at the highest relaxin concentration was considered non-specific binding. Total binding was usually about 7 to 10% of the total counts added and specific binding was about 30 to 50% of the total binding.

That was not bad but further improvements would have been quite tolerable. Human relaxin contains a tyrosine in position A3 which could be selectively iodinated by the same method as described for the RLF-tracer preparation[6] (see also chapter 16). There was really no good reason to expect an improvement from this derivative, and nature agreed. There was also no significant difference between porcine and human relaxin tracer in terms of nonspecific binding.

One usually associates high nonspecific interaction with a high content of basic residues on the ligand surface and that feature, which is shared by all relaxins, would offer a more reasoned approach to the nonspecific binding-problem. Consequently the four basic amino acids of the A chain were replaced by the uncharged citrulline residues. Tetra-citrulline human relaxin retained high potency and acquired an acidic isoelectric point (see chapter 11) but did not decrease significantly the annoying nonspecific binding. The discovery that the loss of its basic amino acids left relaxin quite undisturbed was a generous consolation price.

Scatchard analysis of the relaxin receptor-binding data revealed a monophasic mode of interaction and a 0.5 nM dissociation constant which is well within the range of other peptide hormone-receptor systems. Tissue specificity of relaxin receptor distribution was investigated in the suspected target tissues such as uteri and ovaries of estrogen-primed mice. Brain, liver, kidney and skeletal muscles were used as negative control. While liver, kidney and muscle membranes were clearly negative, crude membranes from mouse brain showed consistently specific binding that was as strong as the binding observed on crude uterine membranes (Fig. 16.3b). The Scatchard plot prepared from brain membrane preparations is similar to the uterus-derived one and so was the number of receptors per mg of protein.[2] Brain can be used without estrogen pretreatment and therefore many of the previously described dose response determinations were performed with crude mouse brain receptor preparations.

At about that time Osheroff and her colleagues published the first of a series of papers on relaxin receptors in rats. They developed a homogeneous ^{32}P-labeled human relaxin tracer with high specific activity (5 Ci/μmol)[7] from recombinant B33-human relaxin which contains an enzymatically phosphorylatable serine residue in position B32. This tracer made it possible to demonstrate clearly on tissue slices the presence of relaxin receptors in the uterus, cervix and brain. Relaxin binding in the brain was found in discrete regions of the olfactory system, neocortex, hypothalamus, hippocampus, thalamus, amygdala, midbrain and medulla, in both, female and male rats (Fig. 17.2).[8]

The idea of short circuit (as opposed to endocrine) action becomes more attractive with a report concerning the distribution of relaxin mRNA in brain.[9] Accordingly, expression of relaxin genes in rat brain is limited to the tenia tecta, pyriform cortex, orbital cortex, neo cortex, ammon's horn, the dentate gyrus and the hippocampus. The relaxin receptors are there and the relaxin genes are expressed in the same regions of the brain; this must have been a fine moment for Gillian and Fred Greenwood, the early proponents of autocrine/paracrine mechanisms for certain relaxin actions.

The brain is only one of the "non-classical" relaxin target tissues. Like the brain, heart tissue of both sexes also binds radioactively labeled relaxin. As shown in Figure 17.3 specific binding was clearly observed in the heart atria. The aorta and other arteries also bound relaxin but here the ligand was not readily displaced by a 1000-fold excess of unlabeled relaxin.[10]

It seemed reasonable to assume that relaxin receptors are regulated differently in reproductive organs than in other relaxin-containing tissues. In Figure 17.4 it is shown that ovariectomy caused a reduction of relaxin receptor in uteri to 53% while supplement of estrogen restored the receptor level to 90%. Testosterone had no effect on relaxin-receptor expression in rat uteri. In contrast the receptor in the heart atrium was not affected by any of these conditions. Relaxin receptors in the heart and the brain of male and female rats have identical affinity constants which in turn were indistinguishable from the affinity of relaxin to the rat uterus (K_D = 1.3 nM).

The first localization of relaxin receptors in non-reproductive tissue, i.e., the rat brain, was not totally unexpected. Brain seems to contain a large number of receptors for hormones that normally would appear only external to the blood brain barrier. The very sharply defined distribution of receptors in distinct regions of the brain left the discoverers, as it would most readers, with the impression that relaxin should have a distinct function in the brain.

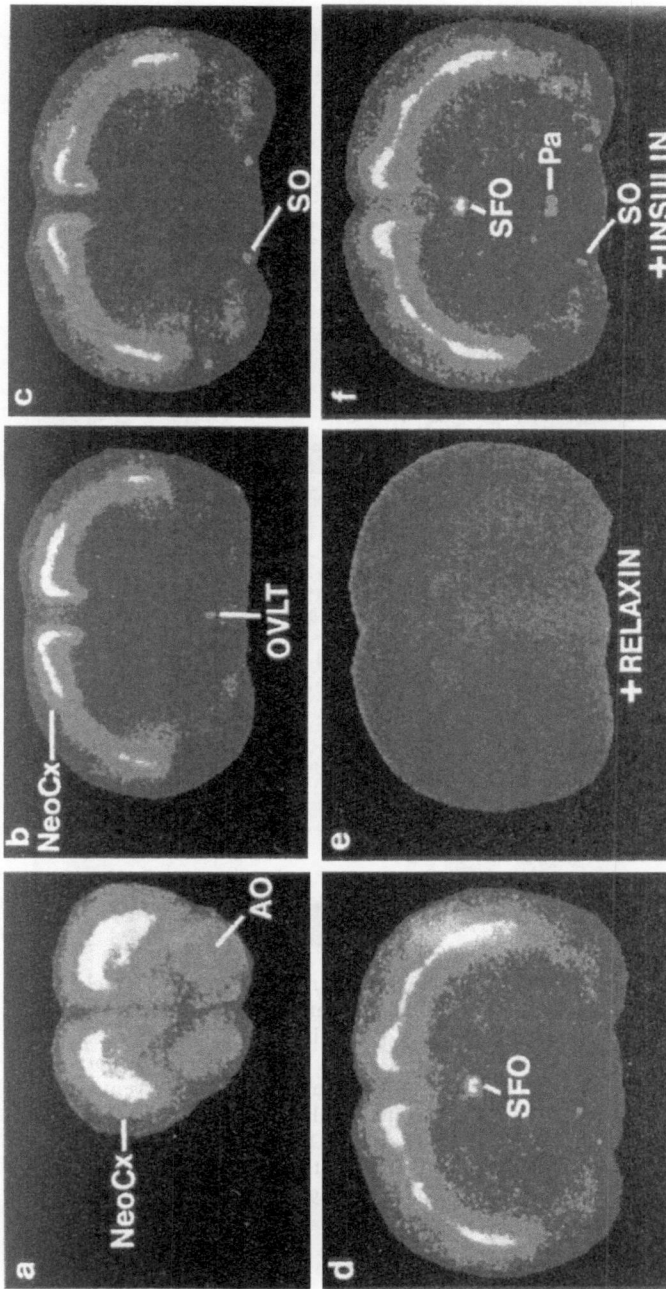

Fig. 17.2. Relaxin receptors in rat brain. A pseudocolor representation of ^{32}P-relaxin binding sites in female rat brains. Low to high binding-intensities are shown in OD units and are represented by a color spectrum from magenta to red (see figure 17.3 and 17.4 for intensity spectra). Panel a: frontal region of the neocortex (NeoCx) and the anterior olfactory nucleus (AO) interact with the relaxin tracer. Panel b: parietal region of the neocortex (NeoCx) shows relaxin-binding sites on the neocortex and the organum vasculosum of laminia terminalis (OVLT). Panel c: binding of ^{32}P-relaxin to the supraoptic nucleus (SO). panel d: binding to the subfornical organ (SFO), panel e shows the lack of binding in the presence of excess unlabeled relaxin. Panel f shows that radioactive relaxin binds to target tissue in the presence of insulin (Pa = paraventricular nucleus). Reprinted with permission from Osheroff PL, Phillips HS. Proc Natl Acad Sci USA 1991; 88:643–6417. (See color insert, page 196.)

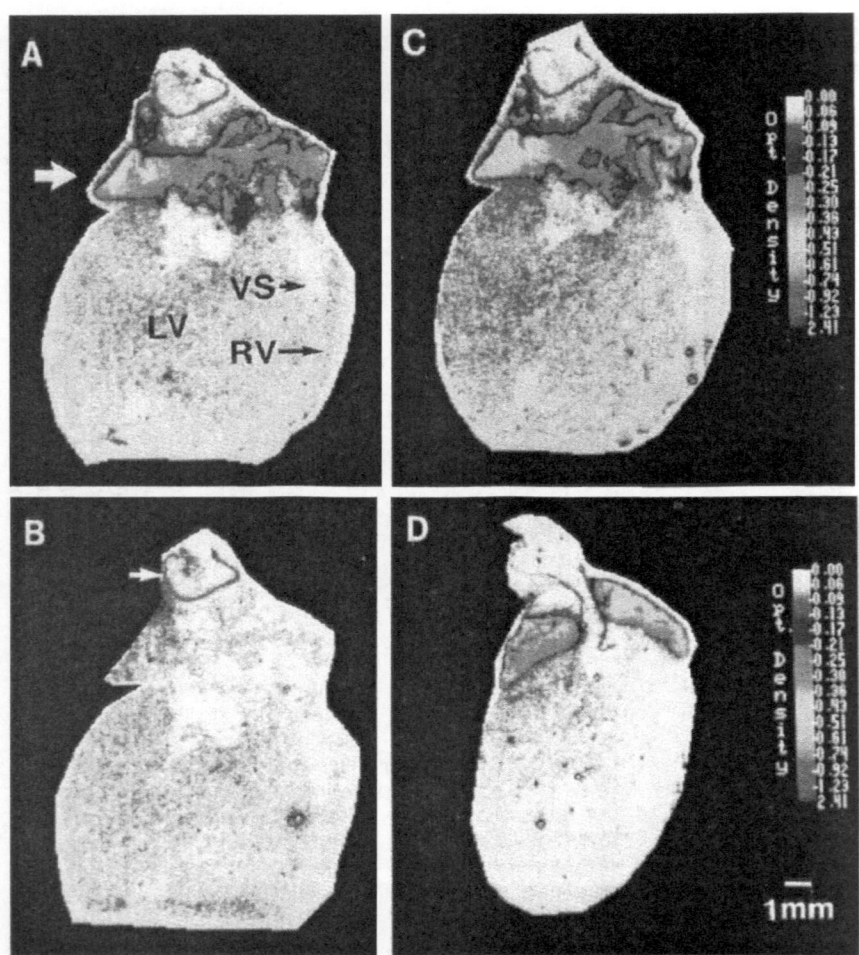

Fig. 17.3. Relaxin receptors in the rat heart. A pseudocolor representation of ^{32}P-relaxin-binding sites in female rat heart (panel A) in the presence of a 1000-fold excess of unlabeled relaxin (panel B) and 1000-fold excess of insulin-like growth factor I (panel C). Relaxi- binding to the male heart is shown in panel D. Tissue sections represent cross sections of the heart. (LV = left ventricals, RV = right ventricles, VS = ventricular septum). The arrow in panel A points to the atrium and the arrow in panel B points to the aorta. Low to high binding-intensities are shown in OD units and are represented by a color spectrum from magenta to red. Reprinted with permission from Osheroff PL, Cronin ML, Lofgren JA. Proc Natl Acad Sci USA 1992; 89:2384-2388. (See color insert, page 197.)

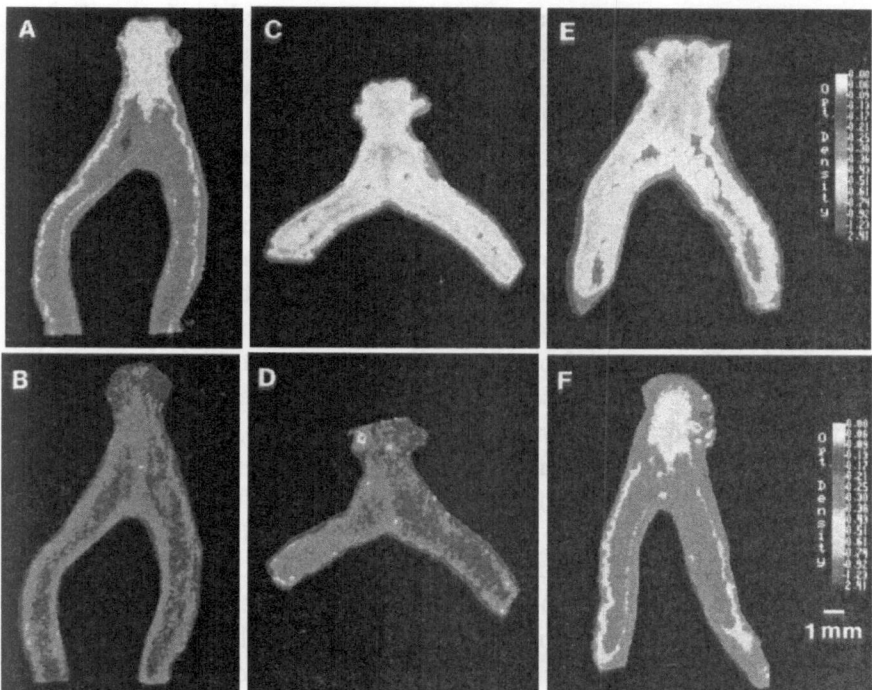

Fig. 17.4. Binding of 100 pM ^{32}P-human relaxin in the uterus of an ovariectomized rat (panel A), an ovariectomized rat but in the presence of 100 nM unlabeled relaxin (panel B), a normal intact rat (panel C), a normal intact rat in the presence of 100 nM unlabeled relaxin (panel D), an ovariectomized rat treated with estrogen (panel E) and an ovariectomized rat treated with testosterone (panel F). Low to high binding-intensities are shown in OD units and are represented by a color spectrum from magenta to red. Reprinted with permission from Osheroff PL, Cronin ML, Lofgren JA. Proc Natl Acad Sci USA 1992; 89:2384-2388. (See color insert, page 198.)

However, the function of relaxin in any non-reproductive tissue is unknown and it must be considered that the brain relaxin receptors may have no function at all. The idea that the mere presence of a protein means that there should be some function for it to fulfill comes from the pan-selectionism concept of the Darwinian hypothesis. Accordingly, for anything to exist in a biological system it must be selected for under evolutionary pressure, meaning a vital need. This may be incorrect but, nonetheless, the convincing demonstration of a well-defined relaxin receptor distribution in the brain will not be lost upon the research community.

Relaxin receptors have also been discovered in the rat heart atrium by the same research group. While it has been possible to produce ionotropic and chronotropic effects in rats with relaxin the question as to the physiological importance still lingers. In order to be ascribed an important cardiovascular role relaxin would have to produce this effect in males as well as females, but the serum relaxin level in males is essentially non-detectable. Over and above political correctness one must allow for gender differences even to the extent of different heart physiology. The fine-tuning of the female circulation to meet the extreme demands of pregnancy may well be in part a relaxin effect exerted on the atria of the heart. Male hearts will bind relaxin as well, but considering the lack of relaxin in male sera this interaction may be of no significance. Again, the cardiac effect could also be exerted by an autocrine or paracrine mechanism which would eliminate the need for circulating relaxin. Either way, the problem has been outlined, it is very interesting and it will certainly attract attention in the near future.

Kuenzi and Sherwood used biotinylated porcine relaxin in their survey of relaxin receptor-expressing tissues in the intact pregnant rat. For localization they did not use avidin but rather gold-conjugated anti-biotin antibodies. Biotinylated relaxin was injected intravenously and the animals were killed after one hour. Tissue sections were flooded with anti-biotin antibodies which marked the relaxin located on its receptor. The gold-labeled antibodies were observed in rather distinct regions in the epithelial cells of the cervix, mammary glands and nipples, as well as in the smooth muscle components of these tissues.[11] To extend this research to larger domestic animals (the pig) an in vitro methodology was developed that allowed the investigation of tissue sections of pregnant and ovariectomized pigs in which pregnancy was maintained by progesterone-treatment. The study confirmed the findings in the rat and in addition revealed that relaxin binds specifically to blood vessels (cervix, mammary glands, nipples and small intestine), to smooth muscles (small intestine), and to skin (back, ear, thigh, and leg).[12]

This observation would suggest that relaxin receptors are widely distributed in the rat and in the pig. Considering how difficult it was to find evidence for relaxin receptors in the first place this abundance was surprising. On the other hand, the observation is in harmony with the observed pharmacological effects of relaxin.

This increasing variety of potential target tissues for relaxin encouraged efforts to generate either primary cell cultures of target tissue or to search for relaxin-receptor bearing cell lines. Homogeneous cell cultures would put another twist into the technical problem of reduction of high nonspecific binding. McMurtry et al cultured fibroblasts from the mouse symphysis pubis but at that time methods were not well enough developed to establish an acceptable relaxin receptor-binding assay.[13] Later, when relaxin acquired commercial interest primary cultures of human uterine cells[14] and rat atrial cardiomyocyes were established.[15] Although this was a step forward, primary cell cultures persist only for a limited number of cell divisions. Immortalized cell lines have distinct advantages but are not readily established. In 1996 the first relaxin receptor-bearing human cell line of the monocyte/macrophage lineage, the THP-1, was described in the literature.[16] Although relaxin-receptor density was low and their number was not modulated by estrogen or progesterone the binding was relaxin-specific. The dissociation constant of 0.1 nM is about an order of magnitude stronger than previously observed relaxin-receptor interactions. This hematopoetic cell line appears to be relatively undifferentiated and the discoverers believe that THP-1 cells are fairly immature monocytes and suggest a potential role of relaxin in macrophage biology.[16]

What is there to be known about receptors that warrants all this effort? The structure of the receptor, and most of all the fine-structure of the ligand binding-site are the essence of physiological signaling. Of course one would also like to know what happens when the ligand hits the binding-site, the mode of signal transduction and propagation within the cell and the chain of reactions elicited by the ligand. Protein chemists need to know the structure of the binding-region of a receptor if the design of ligands and inhibitors is on their minds.

The path from demonstration to the structure of a receptor is tortuous at best, particularly if the receptors are hidden among thousands of others as a small component of a cell membrane. The first attempts to label the receptor were made with either [35]P-relaxin[15,16] or ^{125}I labeled relaxin (our unpublished results). In each case the relaxin was crosslinked to the receptor with a bifunctional crosslinker and the complex extracted with detergents. The receptors were derived from human uterine cells,[15] crude membranes of mouse uteri (our unpublished results), or non-classical relaxin targets such as,

Fig. 17.5. Cross-linking of ^{32}P-relaxin to receptors on THP-1 cells. Cells (500,000 cells/assay) were incubated at room temperature with 400 pM tracer in the presence (R) or absence (T) of 400 nM unlabeled relaxin or 400 nM insulin (I), and crosslinked with EDC [=1-ethyl-3-(3-dimethylamino-propyl)-carbodiimide hydrochloride] or DSS (disuccinimidylsuberate). In a control experiment (–) no crosslinker was added. After

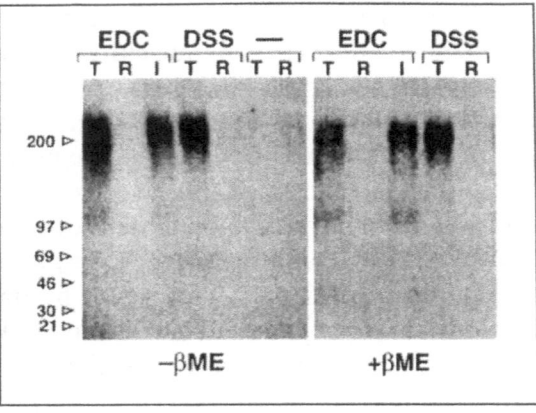

60 minutes the cells were lysed with triton and SDS-containing buffer, heat denatured and separated on a 4-12% SDS-PAGE gel in the presence or absence of β-mercaptoethanol. Dried gels were exposed to Kodak X-Omat film for 20 days. Reprinted with permission from Parcell DA, Mak JY, Amento EP et al. J Biol Chem 1996; 271:27936-27941.

rat myocytes,[15] crude membranes of mouse brain (our unpublished results) and the THP-1 cell line.[16] The receptor size of 220,000 dalton as measured by SDS PAGE appeared to be about the same in each case. A smaller molecule, most likely a degradation product, was observed on some gels which varied from 37,000 Da,[15] to 80,000 Da (our unpublished results) or 100,000 Da (Fig. 17.5).[16] Reduction with β-mercaptoethanol did not reduce the size of these receptors, a feature that makes relaxin receptors different from insulin receptors. The latter is a heterotetramer of the structure $(\alpha\beta)_2$ and a Mr of approximately 350,000 Da which, upon reduction, results in two subunits. Insulin receptor,[17,18] insulin-like growth factor type I receptor[19] and bombyxin receptor[20] are multiple subunit receptors while the relaxin, the insulin-like growth factor type II,[21] and the RLF receptors (our unpublished results) are single subunit structures of approximately the same size.

An important discovery for receptor isolation was that a human monocyte cancer cell line (THP 1) expressed in the membrane an active relaxin receptor.[16] Each cell has, however, only about 1000 receptors so that a billion (10^9) cells would carry 10^{12} receptors approximately, i.e., 1.5 femtomoles of receptor protein. To isolate the protein for initial sequencing requires approximately one picomole of protein. This means substantial, but not insurmountable, upscaling of cell-growing capacity. A promising procedure for the isolation of the relaxin-receptor involves binding of radioactively-labeled relaxin to intact cells, chemical crosslinking of the relaxin-receptor complex, solubilization of the receptor-relaxin complex and separation on lectin columns, i.e., wheatgerm agglutinin, followed by separation by size and by charge. Electrophoresis in SDS-containing buffers with subsequent blotting onto PVDF membranes will provide samples for sequence analysis. The results of these experiments served as a reminder that serious upscaling of the procedures was indeed in order. Nevertheless, there is a path and the problem is essentially reduced to a question of time.

References

1. Cheah SH, Sherwood OD. Target tissues for relaxin in the rat: Tissue distribution of injected ^{125}I-labeled relaxin and tissue changes in adenosine 3'5'-monophosphate levels after in vitro relaxin incubation. Endocrinology 1980; 106:1203-1209.
2. Yang S, Rembiesa B, Büllesbach EE et al. Relaxin receptors in mice: Demonstration of ligand binding in symphyseal tissues and uterine membrane. Endocrinology 1992; 130:179-185.
3. Hock RA, Hollenberg MD. Characterization of the receptor for epidermal growth factor-urogastrone in human placenta membranes. J Biol Chem 1980; 255:10731-10736.
4. Mercado-Simmen RC, Bryant-Greenwood GD, Greenwood FC. Characterization of the Binding of ^{125}I-relaxin to rat uterus. J Biol Chem 1980; 255:3617-3623.
5. Garibay-Tupas JL, Maaskant RA, Greenwood FC et al. Characteristics of the binding of P-32-labelled human relaxins to the human fetal membranes. J Endocrinol 1995; 145:441-448.
6. Büllesbach EE, Schwabe C. A novel Leydig cell cDNA-derived protein is a relaxin-like factor (RLF). J Biol Chem 1995; 370:16011-16015.
7. Osheroff PL, Ling VT, Vandlen RL et al. Preparation of biologically active 32p-labelled human relaxin: Displaceable binding to rat uterus, cervix, and brain. J Biol Chem 1990; 265:9396-9401.
8. Osheroff PL, Phillips HS. Autoradiographic localization of relaxin binding sites in rat brain. Proc Natl Acad Sci USA 1991; 88:6413-6417.
9. Osheroff PL, Ho WH. Expression of relaxin mRNA and relaxin receptors in postnatal and adult rat brains and hearts. Localization and developmental patterns. J Biol Chem 1993; 268:15193-15199.
10. Osheroff PL, Cronin MJ, Lofgren JA. Relaxin binding in the heart atrium. Proc Natl Acad Sci USA 1992; 89:2384-2388.
11. Kuenzi MJ, Sherwood OD. Immunohistochemical localization of specific relaxin-binding cells in the cervix, mammary glands, and nipples of pregnant rats. Endocrinology 1995; 136:1367-1373.
12. Min G, Sherwood OD. Identification of specific relaxin-binding cells in the cervix, mammary glands, nipples, small intestine, and skin of pregnant pigs. Biol Reprod 1996; 55:1243-1252.

13. McMurtry JP, Kwok SCM, Bryant-Greenwood GD. Target tissues for relaxin identified in vitro with ^{125}I-labelled porcine relaxin. J Reprod Fert 1978; 53:209-216.

14. Fei DTW, Cross MC, Lofgren JL et al. Cyclic AMP response to recombinant human relaxin by cultured human endometrial cells—a specific and high throughput in vitro bioassay. Biochem Biophys Res Commun 1990; 170:214-222.

15. Osheroff PL, King KL. Binding and cross-linking of ^{32}P-labeled human relaxin to human uterine cells and primary rat atrial cardiomyocytes. Endocrinology 1995; 136:4377-4381.

16. Parsell DA, Mak JY, Amento EP et al. Relaxin binds to and elicits a response from cells of the human monocytic cell line, Thp-1. J Biol Chem 1996; 271:27936-27941.

17. Ebina Y, Ellis L, Jarnagin K et al. The human insulin receptor cDNA: The structural basis for hormone-activated transmembrane signalling. Cell 1985; 40:747-758.

18. Ullrich A, Bell JR, Chen EY et al. Human insulin receptor and its relationship to the tyrosine kinase family of oncogenes. Nature 1985; 313:756-761.

19. Ullrich A, Gray A, Tam AW et al. Insulin-like growth factor I receptor primary structure: Comparison with insulin receptor suggests structural determinants that define functional specificity. EMBO J 1986; 5:2503-2512.

20. Fullbright G, Lacy ER, Büllesbach EE. The prothoracicotropic hormone bombyxin has specific receptors on insect ovarian cells. Eur J Biochem 1997; 245:774-780.

21. Massaugué J, Czech MP. The subunit structure of two distinct receptors for insulin-like growth factors I and II and their relationship to the insulin receptor. J Biol Chem 1982; 257:5038-5045.

CHAPTER 18

Relaxin as a Drug

Spectacular successes, alternating with equally impressive failures, have kept up expectations and, even in the mind of skeptics, enough conditioned doubts for them to recall relaxin whenever connective tissue causes problems.

The link to relaxin has been established by the observation that chronic symptoms often show complete remission during pregnancy when relaxin levels are relatively high. Endocrinologists, of course, know very well that during pregnancy many hormones change dramatically, but presently the pendulum is swinging the relaxin way. Within the recent past several biotechnology companies have been seeking association with relaxin researchers in order to pursue relaxin and relaxin-like molecules as regards pharmacological action. At least one clinical trial is coming to an end soon enough for our readers to learn about the prospects.

The story of relaxin therapy had its subtle beginning when Wislocki and Streeter made the observation that in rhesus monkeys the blood vessel endothelial cells of the uterine myometrium below the implanted embryo proliferate as if stimulated by a potent growth factor.[1] The term epithelioid cytomorphosis (ECM) was coined by Hisaw who showed that relaxin is primarily responsible for this effect (Fig. 18.1).[2] Raynaud's phenomenon, a peripheral blood flow impairment that causes cold sensitivity of the extremities, completely disappeared during pregnancy when relaxin is produced. Casten and his colleagues made the assumption that relaxin appears in the bloodstream early during pregnancy and that it may be responsible for the beneficial effect on the capillary system. Relaxin shows different and unpredictable patterns of appearance in different animals but as concerns humans, luck was on Casten's side. O'Byrne et al reported

Relaxin and the Fine Structure of Proteins, by Christian Schwabe and Erika E. Büllesbach. © 1998 Springer-Verlag and R.G. Landes Company.

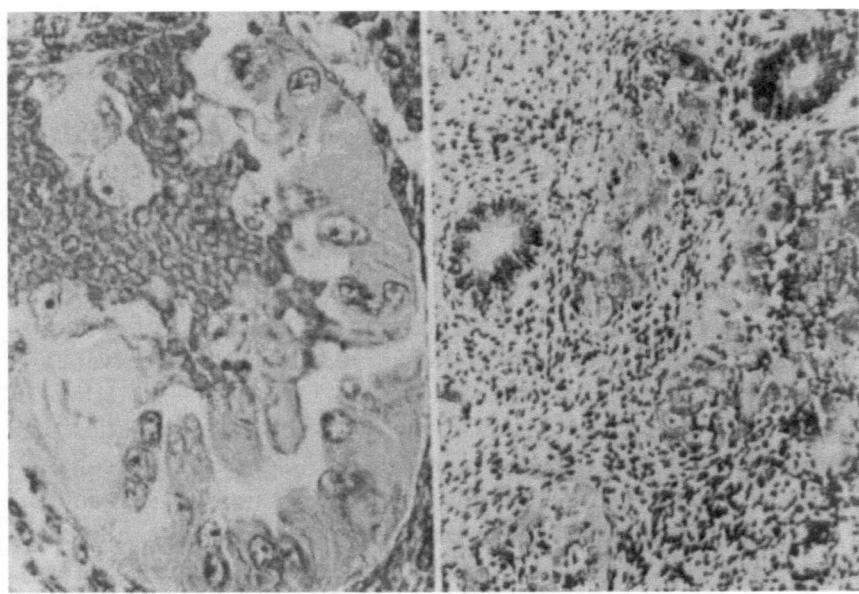

Fig. 18.1. Photomicrograph of endometrial blood vessels of Rhesus monkeys (*Macaca mulatta*) treated with estradiol (10 μg s.c. for 39 days) and relaxin (3000 guinea pig units, s.c., twice daily, for at least 19 days). Left: (x 160) proliferation of endothelial cells (ECM reaction) which almost occlude blood vessels. Right: (x 640) higher magnification of single blood vessel showing hypertrophy and hyperplasia of endothelium. Reprinted with permission from: Hisaw FL, Hisaw FL Jr, Dawson AB. Effects of relaxin on the endothelium of endometrial blood vessels in monkeys (Macaca mulatta). Endocrinology 1967; 81:375-385. © The Endoctine Society.

that relaxin appears in the blood stream of humans as soon as pregnancy can be established with certainty; sows in contrast have elevated relaxin blood levels near term.[3]

The epithelioid cytomorphosis, which according to Steinetz should better be called endothelioid cytomorphosis, and the remissions in Raynaud's phenomenon during pregnancy, again make a logical connection to connective tissue diseases such as scleroderma as well as to peripheral obliterative angiopathies with trophic ulcerations and acronecroses. Rational treatment for chronic connective tissue disorders was not available when Casten and his co-workers took that fateful conceptual leap to introduce relaxin into the clinic on the basis of these findings. They ignored the conventional view of relaxin as a strictly female hormone. Since most of the physiological relaxin effects are dependent upon the presence of estrogen the investigators administered estrogen to all patients, male or female, prior to relaxin treatment. In 1958 they published an extensive paper on the use of relaxin in scleroderma treatment.[4] Patients with scleroderma were studied from 6-30 months; 14 of 23 of these patients had acrosclerotic manifestations. Again, all patients, male or female, were pretreated for two weeks with 1.25 mg Premarin daily.

Relaxin was given twice daily (20 mg of a relatively crude preparation in saline) intramuscularly (i.m.) for one or two weeks followed by 10 mg per day injections in gelatin i.m. Effects of relaxin were observed after 3-5 weeks of treatment.

Trophic ulcers showed marked improvement in 14 out of 18 patients, most of them experiencing complete healing in a relatively short period of time. Raynaud's phenomenon was relieved for 7-8 hours with saline relaxin injections and for 2-3 days if relaxin was given in gelatin.

Of 22 scleroderma patients with general skin tightness 16 reported softening and loosening of skin after treatment. The skin loosening was usually observed first at the site of injection from where it spread over the whole body. Peculiarly the skin of the hands and feet did not respond to relaxin therapy although trophic ulcers healed invariably. Withdrawal of relaxin caused skin tightness to return within 3-10 days, whereas resumption of therapy restored skin loosening. No toxic side effects were observed even after 2-3 years of relaxin therapy.[4]

Casten et al then began to investigate the efficacy of relaxin as treatment of obliterative peripheral arterial disease.[5] Twenty patients were treated at the Miami Heart Institute and 23 at the Vascular Clinic of the Stuyvesant Polyclinic in New York City. In all cases the disease was in an advanced form and conventional therapy had been ineffective. Relaxin was administered to the patients either in a gelatin base at a dose of 20 mg daily or the same amount suspended in oil. All injections were intramuscularly and the therapy was preceded by two weeks of Premarin treatment (1.5 mg orally) three times weekly.

The first study included 43 patients of which 35 had arteriosclerosis obliterans and intermittent claudication; fourteen of the patients were diabetics, and all had lesions from 3 to 26 years duration. Skin temperatures and capillary blood-flow were used as objective criteria to judge the efficacy of relaxin treatment. The most dramatic results again were the "unfailing healing of ischemic ulcers" (Figs. 18.2 and 18.3).[5]

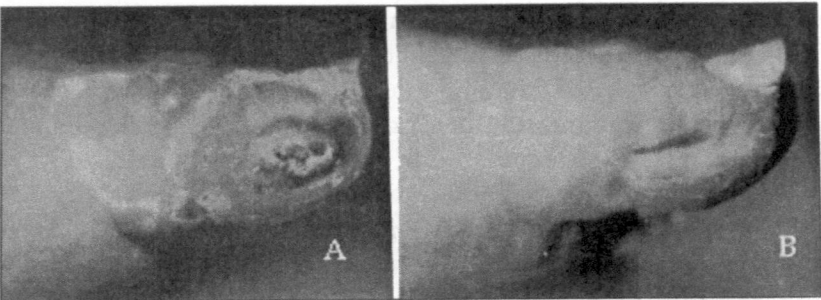

Fig. 18.2. (A) A chronic ulceration of 2 years-duration in an 86 year-old women despite continuous therapy. (B) Ulceration after 4 months relaxin treatment. Reprinted with permission from: Casten GG, Gilmore HR, Houghton FE, Samuels SS. Angiology 1960; 11:408-414.

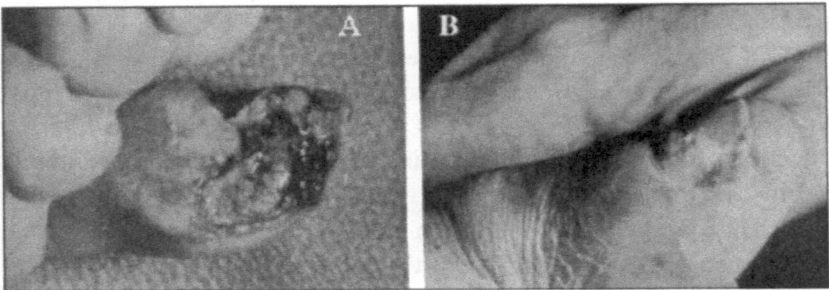

Fig. 18.3. Panel A: Ulceration due to Buerger's disease in a 47 year-old male. The ulcer had been present for 3 months and was extremely painful. Panel B: Same after 3 months of relaxin treatment. Reprinted with permission from: Casten GG, Gilmore HR, Houghton FE, Samuels SS. Angiology 1960; 11:408-414.

The increased skin temperature obtained as a function of relaxin treatment demonstrates that, regardless of pretreatment conditions, the ultimate improvement brought patients back to within one degree of normal (31°C).

A 45 year-old male diabetic had a pretreatment skin temperature of the toes of 27.7°C. The patient received 20 mg relaxin preparation twice a week i.m. which caused the skin temperature to rise to 30°C within one month. Discontinuation of treatment for 2 months caused a drop in skin temperature to 25°C. The treatment was resumed with suppositories (20 mg daily) which caused the temperature to rise again. Skin temperature was maintained at about 29°C for seven months. At month 15 of this study i.m. injections were again performed which caused a further increase of skin temperature (Fig. 18.4).[5]

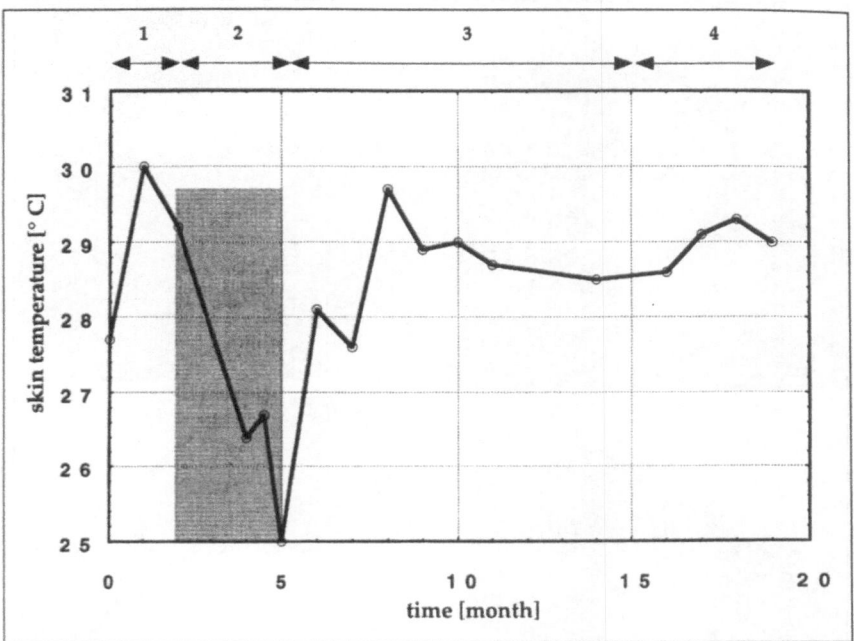

Fig. 18.4. Skin temperature in the toes of a 45 year-old diabetic male. Before treatment the skin temperature was 27.7°C. The following schedule was applied: 1 relaxin i.m. 20 mg twice weekly, 2 treatment was stopped, 3 relaxin suppository 20 mg daily, and 4 relaxin i.m.. Data adopted and reprinted with permission from: Casten GG, Gilmore HR, Houghton FE, Samuels SS. Angiology 1960; 11:408-414.

In lieu of placebos the manufacturer then supplied a low potency relaxin preparation in coded vials which, when given to patients, caused marked decrease in skin temperature within 3 months. When high-potency relaxin was again administered the temperature rose to normal.[5]

Arteriosclerosis obliterans in a 69 year-old Negro caused severe gangrene of the toes as shown in Figure 18.5A,B. After 3 months of relaxin treatment healing of ulcerations and reversal of gangrenous changes are obvious (Fig. 18.5C,D). The skin temperature returned to normal and the ulceration healed completely. At the time these findings were reported the patient had been on relaxin for one year without recurrence of ulceration, retaining normal skin temperature and regaining the ability to walk.[5]

Plethysmographic measurements of segmented and digital blood flow showed increases of 25-32% combined with a 53-110%

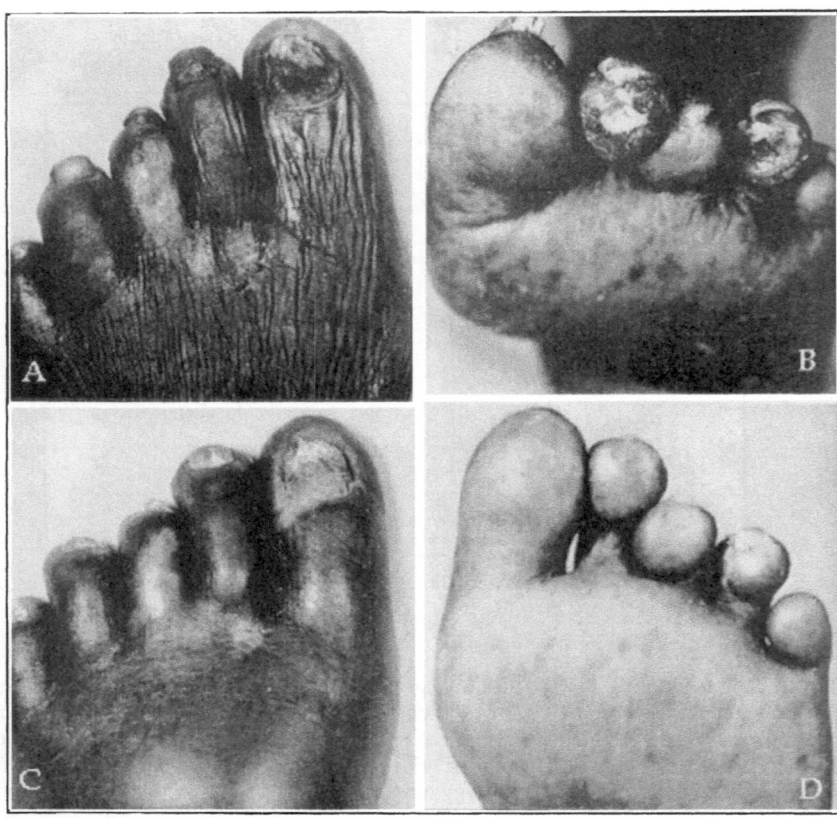

Fig. 18.5. Severe gangrene of the toes of a 69 year-old colored man with arteriosclerosis obliterans. Panel A: affected foot prior to treatment with particular emphasis on the appearance of the skin. Panel B: plantar view of the same toes prior to treatment. The white area appearing on one toe is protruding bone. Panels C and D demonstrate the results of three months treatment with relaxin. Reprinted with permission from: Casten GG, Gilmore HR, Houghton FE, Samuels SS. Angiology 1960; 11:408-414.

increase in pulsation amplitude. Six patients with Raynaud's disease reported a marked increase in cold tolerance.[5]

Relaxin therapy for progressive diffuse scleroderma was studied by Evans. The results were generally affirmative. Relaxin, in combination with diethylstilbestrol and sympathectomy, caused flexibility of the skin in 8 of 11 patients. The effect was most obvious in the face, neck, chest, upper and lower arms, abdomen, and legs. Finger mobility was not improved although ulcers on the hand healed during treatment. One patient presented leg ulcers of 23 years duration that were not influenced by sympathectomy, but healed under relaxin therapy. Subjective improvement of esophageal symptoms was observed in 4 out of 8 patients.[6]

The studies did not elicit the expected response in professional circles. A major critique was the failure of the authors to conduct their studies in a double-blind fashion as required by the FDA. A number of investigators have subsequently tried relaxin in a clinical setting with results being roughly equally divided between success,[7] no conclusion,[8] and failure.[9] Reviewing these studies it became obvious that none of them met the criteria of an unequivocal clinical study, most of them involved far fewer patients than Casten and his colleagues had used, and almost all of them did not use the regimen similar to that used when the original observations were made.

Before the time Casten and his colleagues experimented with relaxin, a great variety of other agents had been tested without success. These agents included steroids, dextran, depolymerizing agents such as penicillamine, chelators such as EDTA in combination with pyridoxine and reserpine, and immunosuppressive drugs such as methotrexate, hydroxyurea, and thioguanine.[10] It appears that relaxin or the crude ovarian relaxin-containing extract has caused probably the most dramatic and most consistent effect. On the basis of literature reports alone one cannot be absolutely positive about the efficacy of relaxin, but it appears that the early results have been promising enough to start a new wave of academic and industrial interest in relaxin and relaxin-like molecules.

Recalculating the dosage of pure relaxin equivalents used in these clinical trials, it would require 1-2 mg of pure porcine relaxin three times weekly to maintain a patient at relaxin levels that produced in the past beneficial effects. It is possible that the aluminum stearate repository form of relaxin[11] might reduce the need for injections to one every two weeks.

In terms of the total units of relaxin given to each patient per injection, a simple calculation reveals an astounding fact. Assuming a body weight of 50 kg for the average patient, the doses used during the clinical studies is 20-40 µg/kg. Physiological experiments are usually performed with 1 µg per mouse which adds up to 50 µg/kg. By any measure it is safe to state that the amounts of relaxin given to patients was not much in excess of physiological concentrations. It is commonly assumed that physiological receptors are much more sensitive to a hormone than the low affinity-type receptors suspected to be in connective tissue or endothelium. True or not, whatever response has been observed had been elicited by very low relaxin levels.

The NIH NPR-l fraction, which was very similar in potency to the Warner Lambert standard preparation, has been purified to homogeneity in the author's laboratory. There were at least 20 fractions other than relaxin in this preparation, but each of the fractions was present in either the same amount as relaxin or less. If a side fraction other than relaxin would be responsible for the connective tissue effect it must be a very potent factor.

Finally one must ask how relaxin would possibly influence connective tissue diseases. It is commonly assumed that specific tissues possess receptors which limit hormonal effects to target organs. It is also accepted that human males do not have an RIA-detectable serum relaxin level, i.e., less than 10 picogram relaxin/ml of blood. In addition, it was thought that relaxin requires estrogen for activity, a steroid normally not prominent in males, which at that point left no rational basis for relaxin action in males. Today it is known that binding of relaxin to certain cell-types is estrogen dependent whereas binding to others is not. The former group of cells is derived from reproductive tissues whereas the estrogen independent cells are not.[12,13]

An answer to this enigma might possibly be found in cell differentiation. The evolution of macroorganisms depends upon cell specialization which may be viewed as selective gene suppression in somatic cells. Thus insulin is produced in islet cells and suppressed in all other cells. From the literature we know that the suppression is incomplete and that, to a small extent, all cells appear to make small amounts of insulin.[14] It is not too unrealistic to expect the same to be true for relaxin receptor messages which might be activated beyond the constitutive level of expression by the required pretreatment of patients with Premarin. This would make a nice story were it not for the fact that the need for estrogen is not established beyond doubt.

This was the state of the art in 1960. Meanwhile the relaxin preparation has changed, from crude porcine relaxin to recombinant human relaxin. Since then the effect on connective tissue has been investigated on a molecular and cellular level in particular by Elaine Unemori and Edward Amento at Connetics Corporation. Accordingly, human relaxin causes significant collagen turnover by stimulating the expression of the metalloproteinase procollagenase, decreasing slightly the level of tissue inhibitor of metalloproteinases, and down-regulating the secretion of interstitial collagens in normal human dermal fibroblasts.[15] The decrease of collagen secretion was observed in untreated fibroblasts but also in fibroblasts in which overexpression of collagen was induced by cytokines, such as transforming growth factor-β and interleukin-1β in combination with indomethacin.[15] This suggested that relaxin may be able to modulate connective tissue turnover in scleroderma where matrix overproduction results in excessive fibrosis of the skin and other organ systems. Indeed fibroblasts derived from the sclerotic skin of scleroderma patients show elevated levels of collagen synthesis and secretion when compared to normal dermal fibroblasts. In this model system human relaxin reduced collagen expression and secretion, indicating its potential beneficial effect in scleroderma patients.[16]

Relaxin's effect on fibrosis in vivo was studied in two rodent models, i.e., the fibrous capsule formation around subcutaneously implanted osmotic pumps in mice and fibrotic infiltration of polyvinyl alcohol sponge implants in rats. The results showed that relaxin caused a decrease in collagen synthesis in vivo under conditions where rapid connective tissue accumulation is occurring. It confirmed the in vitro studies and supported the potential of relaxin in the treatment of pathological conditions of fibrosis.[17] In a subsequent study Unemori et al reported on the effect of human relaxin in the inhibition of lung fibrosis in vitro and in vivo. While human relaxin did not affect basal levels of collagen expression it inhibited the TGF-β mediated over-expression of interstitial collagen by up to 45% in a dose dependent manner.[13]

There is no doubt that relaxin has an effect on connective tissue turnover and that this effect is not restricted to organs of the reproductive system. And yet, the first pharmaceutical application of recombinant human relaxin was directed toward the structural remodeling of the interpubic ligament in preparation for parturition.

A large-scale clinical trial designed to test the efficacy of exogenous human relaxin failed to promote the cervical changes required for parturition over and above normal values for a population.[18] A similar much smaller-scale study suggests that a cream containing porcine relaxin applied to the cervix causes measurable softening.[19,20] Relaxin answers questions not merely with a question but with a puzzle. Is porcine relaxin really more active in humans than human relaxin? Perhaps there is a generic relaxin, one that should be designed according to what structure/function studies have taught us, one that would be better than any of the species-specific ones for this purpose, i.e., a "concept relaxin".

All the signs of endogenous human relaxin activity are observed during parturition.[21] True, symphyseal widening in women is not as dramatic as it is in some other species, but still the overall effect of relaxin on the pubic girdle, including a loosening of the sacroiliac joint, the coccyx, and even the hip joints in extreme cases cannot reasonably be assigned to any other hormonal activity. Why then is exogenous relaxin not effective? The clinical study may have been flawed, and there are voices to that effect, or relaxin action in humans depends on paracrine or autocrine rather than the endocrine mechanisms. For paracrine or autocrine mechanisms the relaxin does not need to circulate but is produced locally and thus independent of the relaxin that is excreted into the blood stream from the corpus luteum of pregnancy. This may explain the observation that women with non-functional ovaries can give normal birth if only the steroids but not relaxin are substituted. The observable effect of porcine relaxin in humans may be attributed to the strange phenomenon that porcine relaxin has a greater affinity for human relaxin receptors than human relaxin. Strange in this case does not mean rare but refers to a result one would not expect in light of our present theories as to how life has assembled itself and evolved (see chapter 19). It is observed that exogenous hormones can be as much as 100 times more effective than the endogenous ones and that effect must be taken into consideration when evaluating a hormone as a drug rather than a physiological agent. In either case, however, we need receptors and the receptors must be connected to the interior of the cell such as to cause or intensify physiological changes.

The statement sounds simple enough but to find a solution has just become even more complex. The difficulty once believed to be solely associated with the problem of specific binding must be extended to include the concept of induced match or stress-producing specific mismatch. As a consequence, binding which is always a precondition is not always sufficient to elicit bioactivity. That case has been made more clearly for rat relaxin than for any other molecule and is discussed in the first chapters in this book. It simply means that a priori the native molecule is not necessarily the best for treatment as opposed to physiological studies, where it is. Pharmacological studies present no limits to the investigators' inventiveness other than efficacy and toxicity.

While new drugs are vitally important for all parties involved and while it is "modern" to find them by random methods there remains forever a part of the human mind that will not rest until the underlying processes are understood. To students one might say that understanding something is a luxury, a privilege, and it is even part of a nations' civility, but aspirin has worked well for humankind long before we knew that it is acetylsalicylic acid and much longer before we knew its mode of action. This does not mean that basic research is unconnected or undirected. It is and must be directed by the researchers' curiosity and only then does it provide the groundwater for a nation from which one can draw as needed, but foremost it provides for the sapience in Homo. If you are troubled finding a fundable purpose for your work you are probably allowing yourself the luxury of doing basic research; keep it up!

As this chapter draws to an end the much anticipated announcement has arrived from the Connetics Corporation in Palo Alto, California. The phase II clinical study involving relaxin suggests that recombinant human relaxin H2 may be effective in the treatment of scleroderma, a life-threatening connective tissue disorder that has been untreatable so far. This is the first controlled clinical study to show positive results.

The double-blind placebo-controlled study involved 64 patients randomized into one of two treatment groups which received 25 µg or 100 µg per kilo per day of either relaxin or placebo. In the group receiving 25 µg per kilo per day statistically significant improvement in skin scores are reported which were the primary clinical end point. In addition positive trends were seen in all other parameters. This is the first ever prospective controlled trial of a potentially disease-modifying agent for scleroderma according to James R. Seibold, M.D., who was in charge of these studies.

Scleroderma is a serious connective tissue disease resulting from an excessive production of collagen. The disease afflicts approximately 300,000 patients in the US with 80% of the cases observed in women. In its severe form, which affects approximately 60,000 patients in the US, the disease results in hardening of the skin and is fatal in 50-70% of the cases within 5 years.

The success of relaxin may be a historical announcement that spells hope for thousands of human beings and, as a little sideline, it means vindication of relaxin as a biologically active, perhaps even uniquely important agent. It also means vindication for two pioneers who had an idea and the conviction to follow through and to risk there reputation trying to proof it correct; Gus G. Casten and Robert J. Boucek.[4]

A very promising pharmaceutical application is based on the relaxin's action on smooth muscle which causes blood vessel dilation and results in blood pressure reduction. Experiments on isolated hearts of rats and guinea pigs indicated a 35 to 40% increase of the coronary flow over normal[22] at a 1 nM concentration of porcine relaxin. In this system relaxin is therefore a 50- to 500-fold more potent blood-flow regulator than acetylcholine or sodium nitroprusside[22] and has a stronger positive chronotropic effect than endothelin, angiotensin II and (-)-isoprenaline.[23] The positive chronotropic effect of relaxin was reported by several authors whereas the positive ionotropic effect is still debated.[22] Relaxin's action on vascular smooth muscle cells parallels the stimulation of the production of nitric oxide, a powerful vasodilator.[22] This observation may explain the significantly lower frequencies of congestive heart failure in cycling women as compared to an age-matched male population; the gender difference disappears after menopause.

The biological response on the isolated rat heart correlates with the presence of high affinity relaxin receptors in heart atria.[12,24] The receptors are gender-independent and appear already one day after birth.[24] When Taylor and Clark found relaxin secretion in male rat atrial cardiocytes the autocrine mechanism became quite plausible. Atrial cells produce relaxin as early as three days after birth which implies that relaxin could be important throughout a lifetime without ever showing up in the circulation, as it is the case in males for example. Most likely relaxin acts on the heart by an autocrine and/or paracrine mechanism rather than by an endocrine loop.[25]

In the rat relaxin receptors an mRNA for preprorelaxin were found in brain. These receptors are gender-independent and are located in regions of the brain known to be associated with blood pressure, fluid balance and release of neuropeptides.[26] The relaxin transcript was observed postnatal at day one as well as in the adult brain in male and female animals, while receptor-binding was first observed at low levels on postnatal day seven in the neocortex. The binding-sites increase during development, reaching the maximum levels of the adult at day 29.[24] The implications have been discussed as well in the chapter on relaxin receptors.

Thus we have reached the place in this chapter where "prospects" should appear in heavy letters that predict miracles to be just around the corner. That is true, of course, except that researchers/authors are the most biased persons on the subject matter at hand.

One could reason that relaxin messages and receptors have been found in all key regulatory locations such as brain, heart and gonads and in widely disseminated tissues such as blood vessels, skin and connective tissue. And that alone makes one think that the next discovery will tell us what their role might be. Furthermore, the technology available today is sophisticated enough to give us a chance at significant discovery in all these vital areas. The baseball player would say that the bases are loaded, all we need is a home run to win big. That is perhaps the most restrained prediction we can offer, but it is not trivial for it may mean a more enjoyable life for very many people who are now facing mere existence. Chronic vascular problems are a nagging affliction not only for the diabetic but for an ever-aging population. Relaxin has been neglected for a long time, and we hope that this chapter helps to impress the reader that this (irrational) neglect has been to our mutual disadvantage.

References

1. Wislocki GB, Streeter GL. On the placentation of the macaque (*Macaca mulatta*), from time of implantation until formation of definitive placenta. Contrib Embryol Carnegie Inst 1938; 27:1-66.
2. Hisaw FL, Hisaw FL, Jr., Dawson AB. Effects of relaxin on the endothelium of endometrial blood vessels in monkeys (Macaca mulatta). Endocrinology 1967; 81:375-385.
3. O'Byrne EM, Carriere BT, Sorensen L et al. Plasma immunoreactive relaxin levels in pregnant and nonpregnant women. J Clin Endocrinol Metabol 1978; 47:1106-1110.
4. Casten GG, Boucek RJ. Use of relaxin in the treatment of scleroderma. J Am Med Assoc 1958; 166:319-324.
5. Casten GG, Gilmore HR, Houghton FE et al. A new approach to the management of obliterative peripheral arterial disease. Angiology 1960; 11:408-414.
6. Evans JA. Relaxin (releasin) therapy in diffuse progressive scleroderma. AMA Arch Dermatol 1959; 79:150-158.
7. Ismay G. Relaxin: Its effect in a case of acrosclerosis. British J Dermatol 1958; 70:171-175.
8. Bradham GB, Stallworth JM, Brailsford LE et al. Clinical evaluation of relaxin in uterine ischemic vascular or collagen diseases. Angiology 1962; 13:418-420.
9. Jefferis JE, Dixon ASJ. Failure of relaxin in the treatment of scleroderma. Ann Rheum Dis 1962; 21:295-297.
10. Birk RE. Treatment of systemic sclerosis. Mod Treat 1966; 3:1287-1301.
11. Anshel J. Relaxin composition and process for preparing same. Patent. US 2964448: 1960:
12. Osheroff PL, Cronin MJ, Lofgren JA. Relaxin binding in the heart atrium. Proc Natl Acad Sci USA 1992; 89:2384-2388.

13. Unemori EN, Pickford LB, Salles AL et al. Relaxin induces an extracellular matrix-degrading phenotype in human lung fibroblasts in vitro and inhibits lung fibrosis in a murine model in vivo. J Clin Invest 1996; 98:2739-2745.

14. Rosenzweig JL, Havrankova J, Lesniak MA et al. Insulin is ubiquitous in extrapancreatic tissues of rats and humans. Proc Natl Acad Sci USA 1980; 77:572-576.

15. Unemori EN, Amento EP. Relaxin modulates synthesis and secretion of procollagenase and collagen by human dermal fibroblasts. J Biol Chem 1990; 265:10681-10685.

16. Unemori EN, Bauer EA, Amento EP. Relaxin alone and in conjunction with interferon-gamma decreases collagen synthesis by cultured human scleroderma fibroblasts. J Invest Dermatol 1992; 99:337-342.

17. Unemori EN, Beck LS, Lee WP et al. Human relaxin decreases collagen accumulation in vivo in two rodent models of fibrosis. J Invest Dermatol 1993; 101:280-285.

18. Brennand JE, Calder AA, Leitch CR et al. Recombinant human relaxin as a cervical ripening agent. Brit J Obst Gynaecol 1997; 104:775-780.

19. MacLennan AH, Green RC, Grant P et al. Ripening of the human cervix and induction of labor with intracervical purified porcine relaxin. Obstetrics & Gynecology 1986; 68:598-601.

20. MacLennan AH, Green RC, Bryant-Greenwood GD et al. Ripening of the human cervix and induction of labour with purified porcine relaxin. Lancet 1980; 1:220-223.

21. Goldsmith LT, Weiss G, Steinetz BG. Relaxin and its role in pregnancy. Endocrinology & Metabolism Clinics of North America 1995; 24:171-186.

22. Bani-Sacchi T, Bigazzi M, Bani D et al. Relaxin-induced increased coronary flow through stimulation of nitric oxide production. Br J Pharmacol 1995; 116:1589-1594.

23. Kakouris H, Eddie LW, Summers RJ. Cardiac effects of relaxin in rats. Lancet 1992; 339:1076-1078.

24. Osheroff PL, Ho WH. Expression of relaxin mRNA and relaxin receptors in postnatal and adult rat brains and hearts. Localization and developmental patterns. J Biol Chem 1993; 268:15193-15199.

25. Taylor MJ, Clark CL. Evidence for a novel source of relaxin: Atrial cardiocytes. J Endocrinol 1994; 143:R5-8.

26. Osheroff PL, Phillips HS. Autoradiographic localization of relaxin binding sites in rat brain. Proc Natl Acad Sci USA 1991; 88:6413-6417.

CHAPTER 19

Relaxin and Genealogy

Relaxin has been a fascinating molecule for the protein chemists, biologists, and for a group of adventurous physicians. Protein designers will be more cautious after learning from relaxin that as little as a methyl group, more or less clear across the molecule away from the binding-site can reduce activity by an order of magnitude. In the context of molecular genealogy, relaxin would have induced humility also in practitioners of this art were they not protected by a lack of receptors. The message nonetheless is that the construed animal genealogy as it is practiced today does not find support in molecular structures as the theory requires.

The single-point origin of life is a Darwinian/neo-Darwinian creed.[1] Life developed from chemistry by an incredible stroke of luck as Jacques Monod, a well known geneticist, put it succinctly. The odds for such an event to occur twice, such that only one genetic code would be visible today, are so small as to warrant no serious consideration. Even one biogenic event based upon chance is said to be legitimized only by the fact that we are here which is, by and by, a false argument. To the contrary, the fact that we are here proves that the model of biogenesis is wrong and that most of the assumptions associated with it are wrong as well. Relaxin has been a catalyst for the development of a new model of evolution, i.e., the genomic potential hypothesis (GPH) which suggests a very different scenario of biogenesis.[2]

What are the consequences of the single-origin assumption, what is conceptually unacceptable about it, how does it clash with evidence and what does relaxin contribute to the story? That is the essence of this chapter.

Molecular evolution as a discipline is predicated upon the assumption that all creatures derive from one start and that also the genes, without exception, stem from one immediately postmordial

Relaxin and the Fine Structure of Proteins, by Christian Schwabe
and Erika E. Büllesbach. © 1998 Springer-Verlag and R.G. Landes Company.

parent gene.[3] The various forms of life today have come about by genomic expansion followed by random mutations, the products of which were sorted out by natural selection according to reproductive success.[1] As a consequence of this philosophy all life forms are by definition either intermediates between a creature of the past and one in the future or the terminal form of a branch. Since we are the product of our genes, our genes must be intermediates in the same way that our bodies, our phenotypes, are if the Darwinists are correct. With nothing but intermediates around it happens that in and outside the discipline of paleontology the intermediates are famous only for their absence, and neither can the intermediate positions of genes be ascertained, in this case, for lack of independent parameters. Still, the idea of intermediates resists well deserved obsolescence as tenaciously as did the phlogistin idea many years back.

In so far as all proteins are contemporary a hypothetical genomic paleo-clock would have to be standardized by comparison with fossil ages, and the genomic changes expressed as the number of mutations accepted per 100 residues must be a linear function of time.[4] If the mutation rates were not uniform in all evolving life forms then the equation linking the number of sequence changes to times of divergence of species has two nonparametric (rate of mutation acceptance and divergence of species) unknowns which is unacceptable to the non-expert.[5] The situation is muddled further by images of converging, diverging, or parallel evolution of gene products. Many of the participants in discussions concerning evolution do not seem to realize these qualities cannot be known unless one assumes the Darwinian theory to be self-evident.

To establish structure/branching time relationships one compares the primary structure of homologous proteins, such as relaxin from species A and B. Figure 19.1 shows all the assumptions and conclusions embodied in the concept of molecular genealogy. The length of the horizontal bar is strictly a function of the sequence differences, i.e., 2% difference is a short bar and 20% difference is represented by one that is 10 times as long. The next dimension needed is the time of branching of the two species (standardizing the clock). Thus one studies the fossil record to determine at what period the two species appeared separately for the first time. When comparing homologous proteins from two mammals, for example, 60 to 100 million years is a fair approximation (the time of the increase in mammalian forms). The protein may have 100 residues and if 10 amino acids are different in the two, than the clock runs at a speed of 10% change per 100 million years. The clock is now standardized. If one finds another pair of homologous proteins from say a bird and a crocodile which show 20% sequence difference then the two branched from each other about 120 million years ago. In Figure 19.2 sequence differences between pig and human insulin are shown graphically. The two hormones differ by one residue (about 2%) and since the fossil record suggests distinct pig progenitors at least 60 million years ago, insulin mutates at a rate of 2% approximately per 100 million years (or 2 PAM units or point mutations accepted per 100 residues and 100 million years). In Figure 19.3 the sequence difference between insulins of humans, pigs, and guinea pigs has been drawn proportional to the numbers used for human and pig. The guinea pig is a mammal with a fossil record as old as that of ungulates, but the sequence differs by 17 residues and the PAMs would therefore be about 34 according to the standardized clock. Assuming the mutation rate is indeed uniform, as we must if molecular genealogy is to be useful, then guinea pigs separated from the common ancestor $1/2 \times 34 \times 100 \times 10^6$, i.e., 1.7 billion years ago! Paleontologists instead maintain that the guinea pigs also have a 60 million year history like other mammals,[6] whereupon the molecular biologists simply replied that guinea pig insulin evolved faster then insulins in other mammals. The disadvantage of this intriguing proposal is that it takes the science out of molecular evolution.

Molecular features seem not to reflect genealogy, regardless of how nice it would have been, but may rather provide some insights into the construction principle of the nuclear core of living systems in general.

There are other conceptual problems that weaken the old idea. Relaxin's major function in mammals, for example, is to make possible parturition of a fairly complex and large fetus by causing dilation of the birth canal. Darwinians therefore consider it an adaptation in the sense that the hormone would have developed under evolutionary survival pressure when placentas evolved about 60 million years ago (their schedule) which would suggest that the relaxin gene did not exist in species older than that. Two of the author's students where curious enough to test this proposition. They obtained ovaries from a 300-pound (150 kg) pregnant sandtiger shark that had been landed by members of the Charleston Shark Fishing Club. When they carried there tissue into the laboratory on both arms

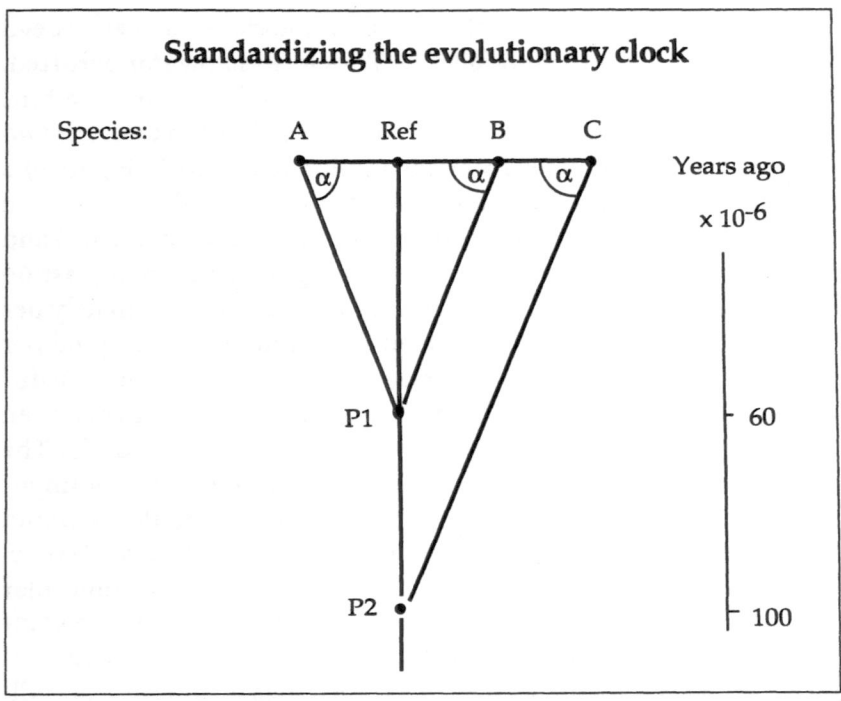

Fig. 19.1. Comparing the primary structures of a homologous protein in species A and a reference species (Ref). The sequence difference is expressed by the distance of the two points on the x-axis. The branching point (P1) on the time-axis is based upon the fossil record of the two species. The resulting triangle has a defined angle α that represents the rate of divergence of the two molecules. The branching of species B and C can now be determined by projecting a line toward the time guided by the same angle (α). Species B has the same number of sequence differences compared to the reference species as does A and therefore the two branched off at the same point (P1) while C with the larger sequence difference branched off about 40 million years earlier (P2).

it seemed quite clear that the rough-hewn deep-sea fisher had played a practical joke on two young medical students, handing them the stomach instead of the gonads. Not with these fellows; they dove into the open body cavity saw pups, one in each uterine tube, and quite sharply concluded that the structure to which the uterine tubes attached were in fact the ovaries. The students managed to extract and purify about 5 mg of the first ever chondrichtian relaxin from several kg of shark ovarian tissue.[7] Although sharks do not have a symphysis pubis the shark hormone knew how to cause widening of the symphysis of a guinea pig and, to a lesser extent, that of a mouse. Relaxin was already fully functional 370 million years ago, when no

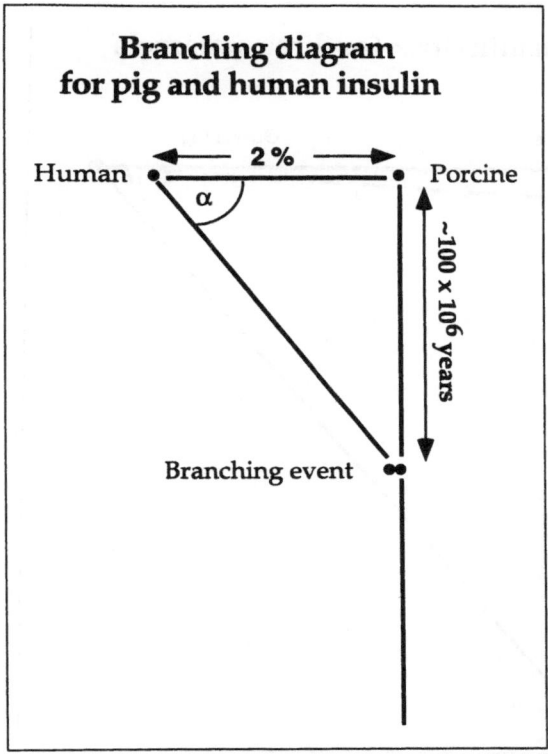

Fig. 19.2. Standardization of the clock with insulin. Porcine and human insulin differ by one amino acid. The active hormone consists of 51 amino acid residues and the difference is therefore 2%. The fossil record indicates that pigs were present about 60×10^6 to 100×10^6 years ago.

traces of mammals were deposited in the fossil record. The idea of adaptational development under evolutionary pressure will just not stand up to the information pressure of modern research. Cases like this have been observed before but now as then the Darwinian mind has mutated itself out of the dire consequence of such controversy. The press, however, showed a keener mind and the story went through main papers here and to the illustrated periodicals like the *Spiegel* and *Geo* in Europe.[8,9] How did the shark get relaxin was the main theme and it is with great pleasure that the conversations with the *Spiegel* staff are remembered for quick-wittedness and intense curiosity that is so rare among peers. The Darwinians recovered fast and all editorial offices were henceforth closed to genomic potentialists.

The relaxin from human, dog, hamster, rat and shark all differ from each other by more than 50% of their respective sequences. This was as disconcerting to the Darwinians as it was normal to the genomic potentialists. The Darwinians were in a severe bind because the insulin of all the mammalian species (guinea pig exempted) are

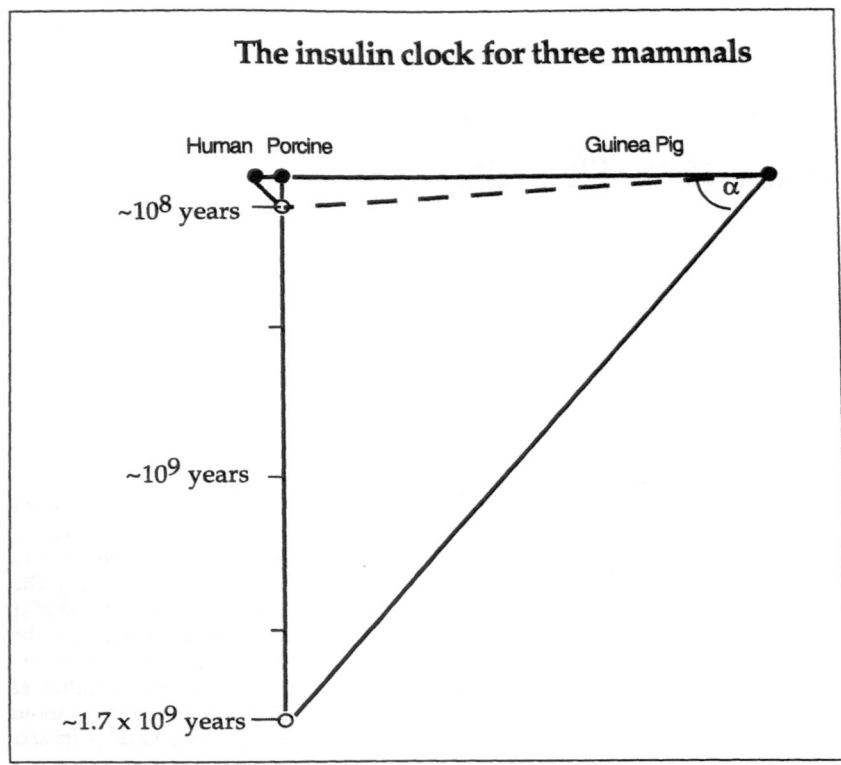

The insulin clock for three mammals

Fig. 19.3. Application of the standardized molecular clock (derived in Fig. 19.2) to guinea pig insulin. Using a constant α the guinea pig would have emerged 1.7 billion years ago. The fossil record suggests that guinea pigs appeared at the same time as other mammals (appr. 100 million years ago). To compensate for this discrepancy it is assumed that guinea pig insulin evolved faster and that rate is represented by the dashed line.

essentially identical. The two proteins, relaxin and insulin, would reflect completely different genealogies, leaving one no rational way to decide between them. The different patterns of evolution presented by these two proteins is *sensu strictu* enough to destroy the neo-Darwinian concept of molecular genealogy.

An even more remarkable instance of support for the GPH comes from murine relaxin. Mouse and rat have always been given juxtaposition in evolution that would include a suggestion of very recent divergence. The mouse relaxin sequence reported as cDNA structure however shows the most dramatic structural change of any of them in the major disulfide ring which is extended by one tyrosine in the penultimate position before the C-terminal cysteine of the A chain. Ordinarily this would be ascribed to an insertion mutation followed by selection so that today one will only find the "large loop-relaxin" in mice (Fig. 14.1). In fact the mouse gene also encodes one variant relaxin with an isoleucine instead of a valine in position A13. Both of these relaxins have the same unusual crosslinking pattern.[10] Here then a concise test offered itself for a Darwinian conjecture. One could synthesize the native mouse relaxin and the hypothetical revertant and compare both in the homologous system. The result of this rather extensive experiment showed a mouse relaxin that was significantly less active than the synthetic des-tyrosine (revertant) relaxin. From both, the bioassay and the receptor-binding assay,[11] it was clear that the hypothetical revertant was a better relaxin in the mouse system. How did the mouse select for the inferior gene when it purportedly split from the rat which has a normally crosslinked and very potent relaxin?

The genomic potential hypothesis predicts that primordial pregenes were distributed by proximity in the chemical pool of developing potential species and that therefore certain sequences could occur in quite separate taxa. The standoff was resolved when a lyophilized ovarian preparation from a baleen whale arrived at our laboratory and when large amounts of material were detected that reacted with our anti-porcine relaxin antibody. Isolation, purification and sequence analysis showed Arg-Met-Thr-Leu-Ser— and so on, the whole porcine relaxin A chain. In a laboratory where porcine relaxin was used in large amounts, contamination comes to mind rather immediately. Even faster were the smart remarks from co-workers that coaxed the experimenter into offering a pizza party bet to the skeptics. A new extraction under surgically clean conditions again yielded the porcine A chain. Contamination seemed no longer plausible. Analogous to the porcine molecule, the whale relaxin B chain began with cyclic glutamic acid followed by arginine (first mistaken for a serine) which was different from the porcine sequence. The whale B chain continued like porcine relaxin until residue 6 which was leucine instead of phenylalanine. Thereafter the identity continued to the end of the B chain for *Balaenoptera edeni* (Bryde's whale). *Balaenoptera acutorostrata* (Minke whale) showed two additional differences so that the Bryde's whale relaxin is closer to porcine relaxin than to its fellow whale.[12] Conversely, porpoise (*Phocoenoides dalli*, Dall's porpoise, a toothed whale) relaxin is again identical to *B. edeni*.[13] Porcine relaxin does not convert cleanly from the prohormone to the active form as insulin does, for example. Not only did the whale have the same relaxin in about the same high amounts but all the conversion side-products were there as well. The skeptics threw a pizza party in good spirit to celebrate the discovery calling seriously into question the author's earlier statement about the absence of earthly rewards in science.

When the sequence data of all known relaxins were reviewed several startling observations could be made.[14] The hormone which is identified not only by biological activity but also by the structure of its binding-site differs by about 55% in animals of purportedly recent divergence and by about the same amount from the relaxin of chondrichtians. By the rules of molecular evolution this would indicate that chondrichtians and mammals diverged at the same time although the fossil record puts a 300 million year distance between them (Fig. 19.4).

What are we to make of it? Relaxins among land mammals are very different but one of them, the pig, has the relaxin of whales. To complicate matters further all sea mammals examined so far have relaxins that are as much alike as insulin is among land mammals. All this seems to be in harmony with the clonal development concept of the new hypothesis (GPH) and contrary to the succession idea of Darwin. The GPH would also predict that other modern relaxin genes such as the human, for example, would be found in simple species. The idea of development in sets has already been advanced by Denton in his excellent and therefore largely ignored book "Evolution, a Theory in Crisis".[15] Perhaps a title like "Darwinism, a Theory in Crisis" would have been more accurate and more acceptable, for evolution in the broadest sense is not in question.

The recent discovery of a relaxin gene in tunicate gonads by Dr. Danielle Georges (UFR de Biologie-Université Joseph Fourier-Grenoble) provides support for these ideas. The work is still in progress at the time of this writing, but the cDNA has been found in her laboratory while we have, in parallel, found the protein in a gonadal extract of tunicates. At this point there is no doubt that tunicates have porcine relaxin! The first records of tunicates date back to 500 million years. Of course, the animals used are contemporary but in as much as their appearance has not changed it is predicted by the GPH that the proteins have not changed much either. The old idea of evolution by adaptation is unable to deal with this evidence. It would not seem reasonable to draw the intermediates between a

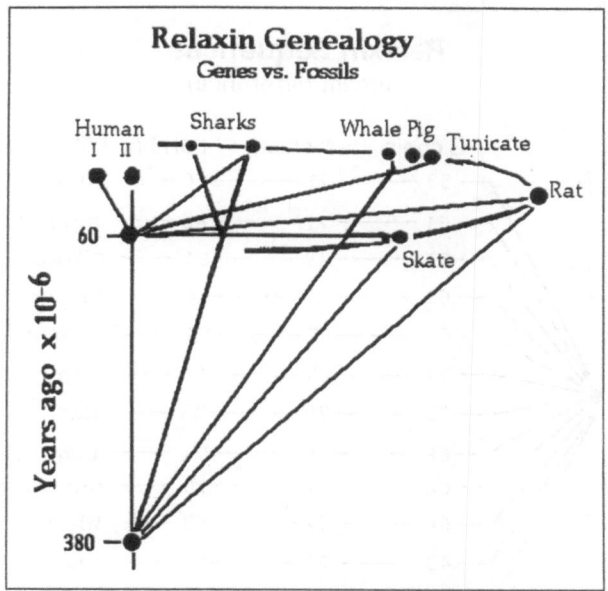

Fig. 19.4. Evolution of relaxin based on human relaxin II as reference sequence. Standardizing the clock on porcine relaxin, for instance, will imply a branching point for all species at about 60 to 100 million years ago but when the clock is standardized on shark relaxin the branching point for all species would be 380 million years ago.

tunicate and a pig and then revert to a whale or to construct a common ancestry via this route. Nor is it plausible to say that relaxin in the pig evolved to its present state via evolutionary pressure in the pig when the gene had already been present at the tunicate stage. Even skeptics will allow that the evolutionary pressure for the two species would have to have been so different that the same outcome, i.e., pig relaxin, would not come to mind.

With all the abuse that comes to authors of new ideas, there comes, off and on, a deeply satisfying moment.

Relaxin is an obvious, but not the only, exception to an imagined orderly progression of succession of species from the Cambrian to our time. Cytochrome and its variants, the celebrated examples of molecular evolution, do not conform to the paradigm when viewed from neutral ground. Consider, for example the diagram comparing relaxin and cytochrome sequences respectively from different species (Fig. 19.5).[5]

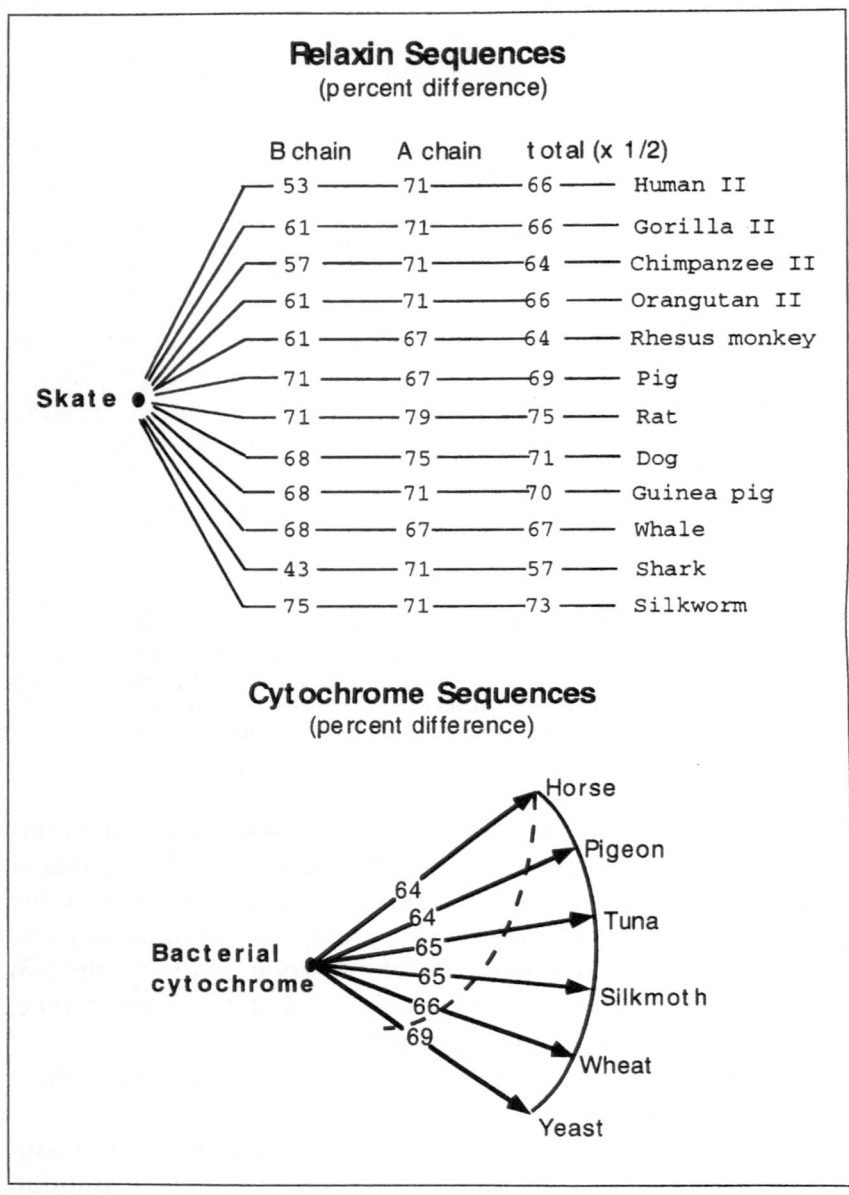

Fig. 19.5. Comparison of molecular differences in relaxin and cytochrome. Skate relaxin and bacterial cytochrome were used as reference. In both cases no correlation is observed between the molecular diversity and the purported phylogenetic rank of species and phyla. The dashed line for cytochrome shows how the molecular diversity should have appeared if to the neo-Darwinian model were correct. The solid line is in harmony with the GPH.

What kind of picture is unveiled by relaxin's refusal to get in line? Within the framework of this book only a brief expedition into the origin of life is permissible, but a glance at the start of living systems is necessary to explain discrepancies between observation and the Darwinian ideas, and how the GPH is different.

Chemistry demands that nucleic acid polymers are the beginning of every life scenario. Still, within the old school the probability of a biogenic event is so low that only at one point one aggregate could have been sufficiently perfected to seed with life all of the earth from the mafic rock to the lower layers of the atmosphere. The new idea, in contrast, points out that chemistry is a mass action phenomenon and that the natural outcome of equilibrium chemistry is never measured in single numbers but rather in numbers of moles, i.e., 6.023×10^{23} molecules. In terms of nucleotides this would amount to about 250 to 300 gram/mole. One must think of perhaps 10,000 moles or 2.5 to 3 tons, or maybe millions of tons, of nucleic acid material to have been produced during the abiotic chemical phase lasting at least one hundred million years. Considering the fact that about 500 million tons of N_2 are falling onto the earth surface every year the assumption of 3 tons of nucleotides is extremely conservative. Why should one care? Because the amount of nucleic acid material on early earth would give one an estimate about the amount of potential biogenic information available. According to the genomic potential hypothesis there was enough to spell out every form of life extinct and extant thousands of times over and over again so that, what we view today as gene duplication and mutations, was indeed the result of reiterative chemistry during the prebiotic period. But this estimate, would it even be high enough to satisfy the hypothesis which purports that the largest amount of nucleic acid material existed at the origin of life? A string made of 10^6 moles of nucleotides would stretch about 100,000 light years clear across the Milky Way galaxy. The length of a human genome (3×10^9 bases) is merely measured in centimeters! The skeptic, tight-rope walking one triple-codon at a time along the transgalactic string of DNA, would probably have seen enough to account for any and every genomic configuration that has been on earth since the beginning of life by the time it reaches Uranus at the edge of our solar system and there is a lot yet to go! The reader must cut loose from the limitation of the world of our experience which has curtailed imagination (in the sense that some phenomena described by the theory of relativity are by and large outside our experience). This string of nucleic acids has the potential to code for so many proteins or nucleic acid structures that there is nothing left to chance. Luck has been banned to the sidelines in this greatest of all plays that terminated in self-consciousness; chemistry has bought all the lottery tickets, winners and losers. The losers referred to here are the short-lived ones that never leave a fossil record, as opposed to extinct species that where successful for millions of years.

Collapsing the giant strand of nucleic acid back onto the earth's surface and cutting it into pieces that are long enough to represent the overall sequence and composition of the original, one could fill millions of water-containing depressions on early earth with nucleic acids to a fairly high concentration. This would be the condition that chemistry would create and it would be the starting point for an immense number of biogenic events that would show a great deal of similarity in very basic functions such as protein synthesis and energy metabolism, and harbor at the same time a great deal of diversity of the more peripheral functions that regulate shape, size, propulsion, and so on. The first prediction of the genomic potential hypothesis is that a large number of organisms should be observed at the moment chemistry crosses the threshold to life. This appears to be supported by evidence. So far about 3,000 fossils of microorganisms have been found in 3.5 billion year-old rocks of which 300 represent different species and 90 of these 300 have modern look alikes.[3]

The multiplicity of origins argument is even more persuasive when one takes a look at the other great divide in biological development, the Cambrian period of macroorganismic evolution. One slab of the Burgess Shale shows a tremendous variety of animals that lived together and died together, and so far more than 120 genera and 30 different major forms have been identified, roughly half of which are not readily assigned to any living taxon.[16] It has been argued for years that the mechanism of gene duplication and mutation would be much too slow to produce such a sudden appearance of variety, and so it happens that one finds here in the Burgess Shale the old concepts of monophyletic evolution buried with all the Cambrian animals.[2,5]

The mosaic distribution of protein structures is a consequence of a development of sets and variations of nucleic acids produced by the complementarity in nucleotide chemistry so that eventually normal distributions of relaxin-like activity or insulin-like activity or hemoglobin-like structures would be produced (Fig. 19.6). During biogenesis different portions of these polymers would be included in the developing cell because of local availability. The sum of the genomic material included within a cell must provide for all the functions that will be required within a few minutes of closure of the cell wall, i.e., the production of a non-equilibrium chemical system that needs to import energy or energy-producing compounds in order to feed its entropy requirements. If this is not achieved, the unit will fall apart again and the components may become part of another cell. The process might be viewed as a cell formation cycle at the interphase of equilibrium and non-equilibrium chemistry (Fig. 19.7).

Cell formation is as much a conceptual bottleneck as it is a chemical one. Because non-equilibrium chemistry will stall without energy influx, it is imperative that many of the maintenance functions such as protein synthesis, and with it energy coupling, had been developed before cell formation could have occurred and before cells could have endured. The protein-coding machinery is almost universal because it was a precellular event and not, as commonly assumed, because it came from a single surviving cell.

Since the configuration for various taxa is laid down at this level of biogenesis one can expect to find the same relaxin in tunicates, whales and pigs, different relaxin in pigs, dogs, humans, rats, mice and the same insulin in humans, pigs, dogs, cows and whales; one just cannot predict it and that is the hint at a different underlying order that relaxin has given us.

A prediction that can be extracted from the Genomic Potential Hypothesis is that the complexity of life one might find on another habitable planet should be proportional to the amount of the abiotically synthesized nucleic acids reservoir. This would mean that half the amount of nucleic acid material on earth may (just for illustration) terminate complexity at the saurian stage, and 10% may limit development to the stage of single cells. Or would someone venture to suggest that humans are the magical stop signal that must be reached regardless of physical reality and that cannot be surpassed; the biological equivalent of the speed of light that limits the physics of the universe?

As the reader begins to lift his eyes off these pages to let them stray through the window out across the landscape, he at once makes the reassuring discovery that nothing has changed! If hypotheses do

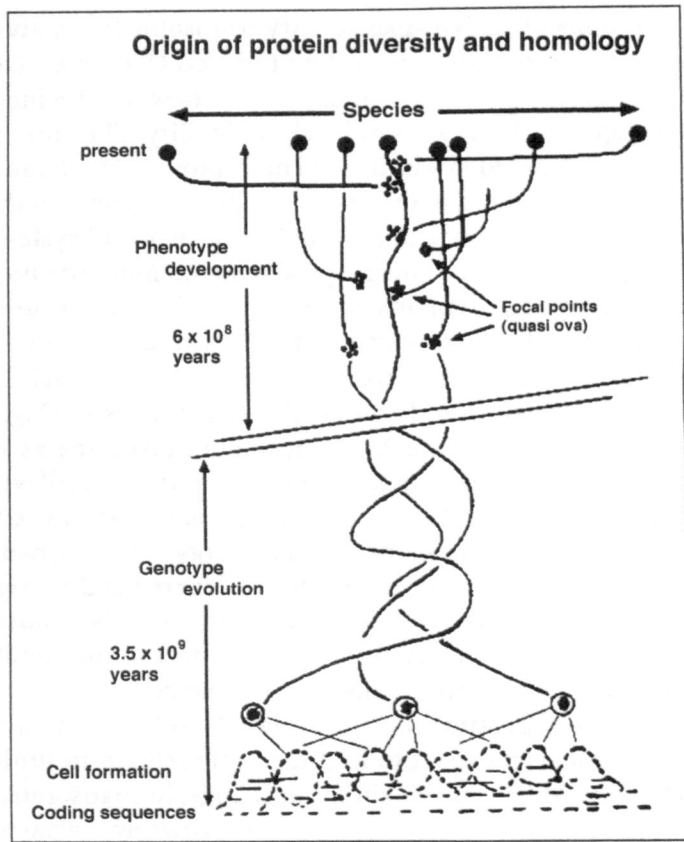

Fig. 19.6. The GPH interpretation of protein diversity and homology. Cells are assembled more or less as sets rather than individuals. In the process they take up a large excess of potential coding material. When a small fraction of the assimilated nucleic acid contains enough information to maintain non-equilibrium chemistry and to self-duplicate, they will tend to persist. The curves are the "worldlines" of the various clones and the time between the origin and the formation of "focal points" or quasi ova from which macroorganisms developed. Each focus again has it's own worldline which is contiguous through the macroorganismic phase transition and has each its own beginning and end. The GPH presents a unified picture that integrates the distribution of protein structures and the fact that species appear in the fossil record rapidly and simultaneously with no hint of intermediate forms.

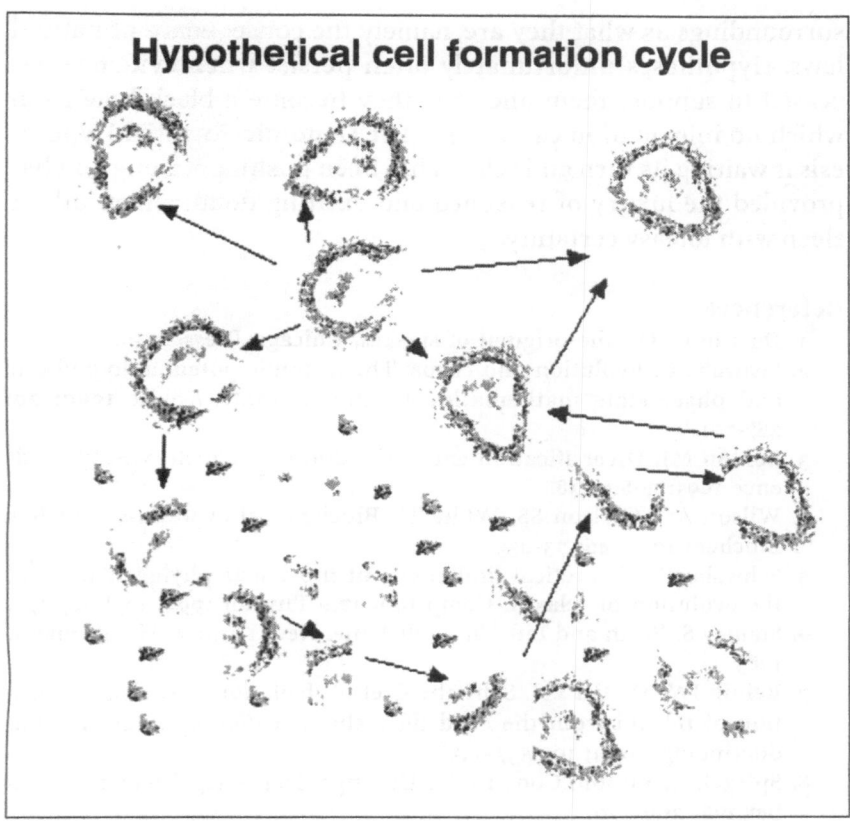

Hypothetical cell formation cycle

Fig. 19.7. The hypothetical cell formation cycle is an attempt to describe how a complex structure can come about and transit to the energy-dependent state within the time frame of minutes at the most. Catalytic particles actively synthesizing proteins and other macromolecules assembled into osmotically active cells. Containment limits diffusion and thereby causes rapid increases in concentrations of all catalysts. If enough functions are represented within the nascent structure the logarithmic increase in protein concentration will spark the condition of life. Insufficient information or absence of only one factor will cause the cell to break up again into fragments that may now associate with others which may or may not contain the complementary functions needed for success.

not change nature, why are we fighting so hard about them? Because they change the way we experience the world and respond to it. A comet today is a large rock or lump of ice from the Ort cloud that buzzes our sun in an eccentric orbit; not too long ago it was clearly the sign of divine anger that needed to be calmed by sacrificing someone else. Many a virgin in as many societies would have enjoyed the intellectual progress had it occurred in due time. Hypotheses guide our effort to understand nature, they provide the impetus for experimentation, they train our senses to recognize features in our surroundings as what they are, namely the consequence of natural laws. Hypotheses unfortunately often persist when evidence has ceased to support them and thus they become a black hole from which no information can escape. The Genomic Potential Hypothesis is waiting its turn and relaxin has been pushing it along and has provided the luxury of reasoned and exciting doubt where others sleep with uneasy certainty.

References

1. Darwin C. On the original of species. Chicago Press, 1859.
2. Schwabe C. Evolution and chaos: The genomic potential hypothesis and phase-state mathematics. Computer Math Applic 1990; 20: 287-301.
3. Benton MJ. Diversification and extinction in the history of life. Science 1995; 268:52-58.
4. Wilson AC, Carlson SS, White TJ. Biochemical evolution. Ann Rev Biochem 1977; 46:573-639.
5. Schwabe C. Theoretical limitations of molecular phylogenetics and the evolution of relaxin. Comp Biochem Physiol 1994; 107B:167-177.
6. Stanley S. Earth and Life Through Time. New York: W.H. Freemann, 1985.
7. Reinig JW, Daniel LN, Schwabe C et al. Isolation and characterization of relaxin from the sand tiger shark *(Odontaspis taurus)*. Endocrinology 1981; 109:537-543.
8. Spiegel. Evolution: Code in der Ursuppe. Der Spiegel 1983 24. October 1983:269-270.
9. Geo. Darwinismus: Der Irrtum des Jahrhunderts. Geo 1984 25. June 1984:75-112.
10. Evans BA, John M, Fowler KJ et al. The mouse relaxin gene: Nucleotide sequence and expression. J Mol Endocrinol 1993; 10:15-23.
11. Büllesbach EE, Schwabe C. Mouse relaxin: Synthesis and biological activity of the first relaxin with an unusual crosslinking pattern. Biochem Biophys Res Commun 1993; 196:311-319.
12. Schwabe C, Büllesbach EE, Heyn H et al. Cetacean Relaxin: Isolation and Sequence of Relaxins from *Balaenoptera acutorostrata* and *Balaenoptera edeni*. J Biol Chem 1989; 264:940-943.
13. Woods AS, Cotter RJ, Yoshioka M et al. Enzymatic digestion on the sample foil as a method for sequence determination by plasma desorption mass spectrometry: The primary structure of porpoise relaxin. Int J Mass Spectrometry Ion Processes 1991; 111:77-88.
14. Schwabe C, Büllesbach EE. Relaxin: Structures, functions, promises and non-evolution. FASEB J 1994; 8:1152-1160.
15. Denton M. Evolution: A Theory in Crises. Adler & Adler, 1985.
16. Conway Morris S. Burgess shale faunas and the Cambrian explosion. Science 1989; 246:339-346.

Color Figures

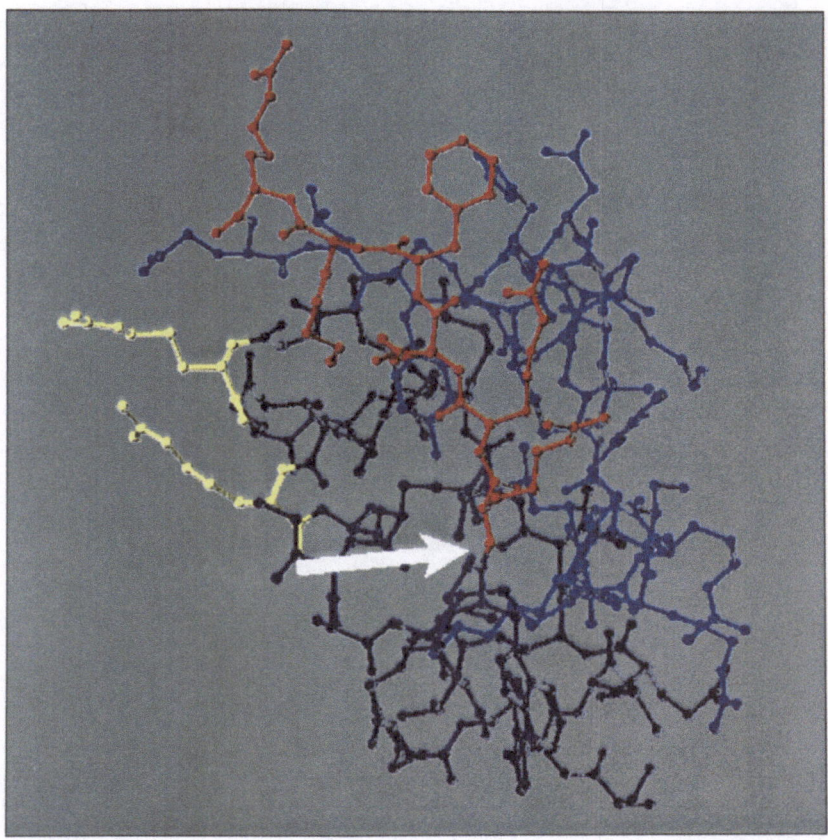

Fig. 8.6. X-ray structure of human relaxin showing the A chain in cyan, the B chain in blue, arginine residues on the B chain helix in white and the artificial connecting peptide in red.

(A, above) Introduction of the peptide Arg-Arg-Glu-Phe-Lys-Arg (red) causes a shift of the N-terminal region of the A chain and the C-terminal region of the B chain but leaves the core structure of the molecule unchanged. The arrow points to the junction of the C terminus of the B chain and the connecting peptide.

(B, see next page) One monomer of the relaxin dimer. (The X-ray data were derived from the Brookhaven databank where they were deposited by Eigenbrot et al 1991). The arrow indicates the point of movement of the C terminus of the B chain in the single chain molecule.

Relaxin, by Christian Schwabe and Erika E. Büllesbach.
© 1998 Springer-Verlag and R.G. Landes Company.

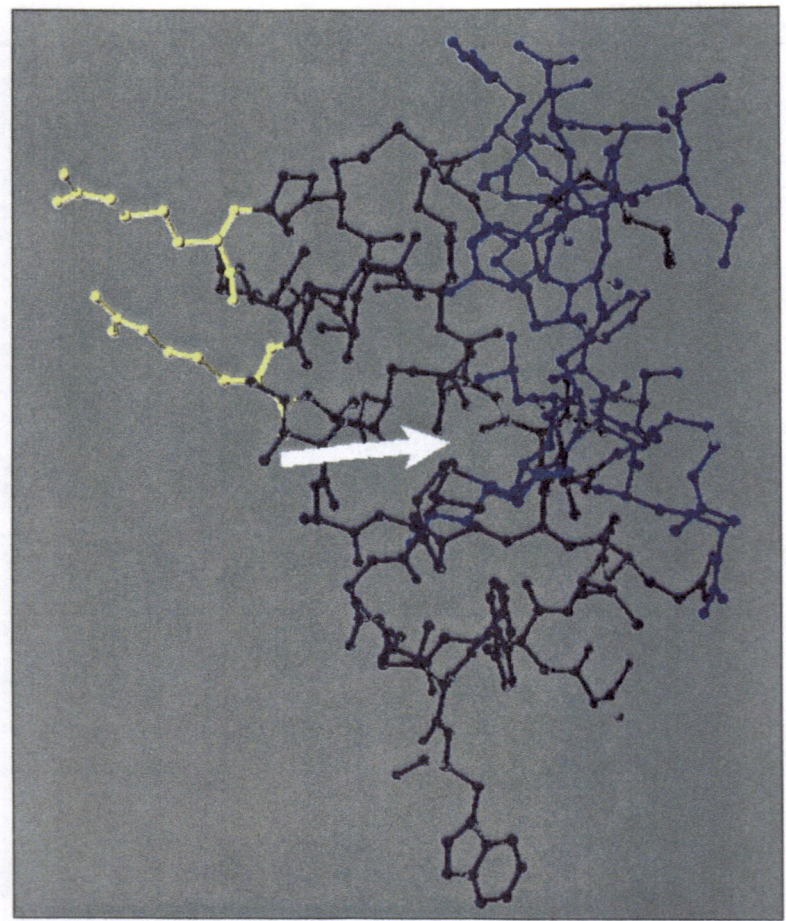

Fig. 8.6B. One monomer of the relaxin dimer. (The X-ray data were derived from the Brookhaven databank where they were deposited by Eigenbrot et al 1991). The arrow indicates the point of movement of the C terminus of the B chain in the single chain molecule.

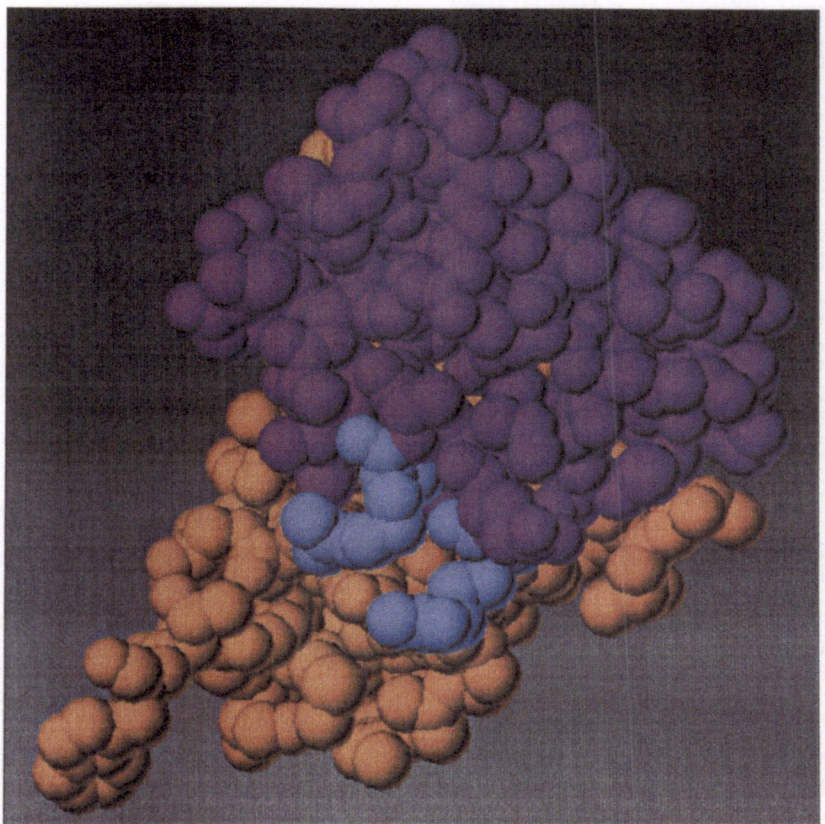

Fig. 10.3. Proposed relaxin receptor interaction site involves the two arginines on the B chain helix (The X-ray data were derived from the Brookhaven databank, where they were deposited by Eigenbrot et al 1991).

(A) Three-dimensional structure of the relaxin dimer. The two monomers are shown in gold and purple, respectively. The proposed active site arginines are shown from one monomer only (cyan), protruding from the B chain helix and stabilized by residues on the surface of the second monomer.

(B, on following page).

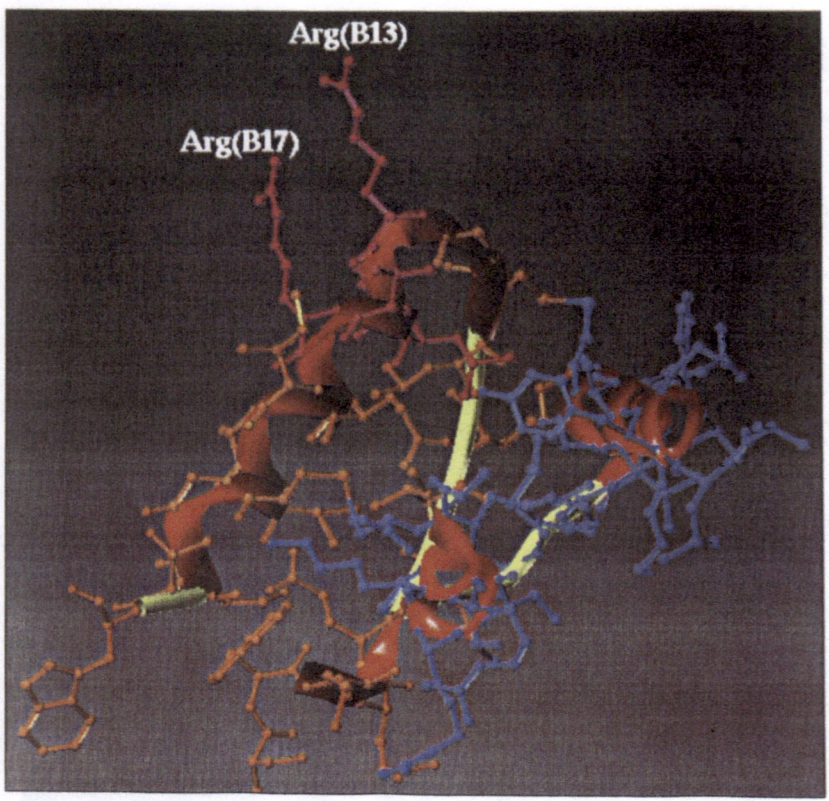

Fig. 10.3. (B) Secondary structure of one monomer with arginines pointing into the water. The A chain is shown in cyan and the B chain in gold.

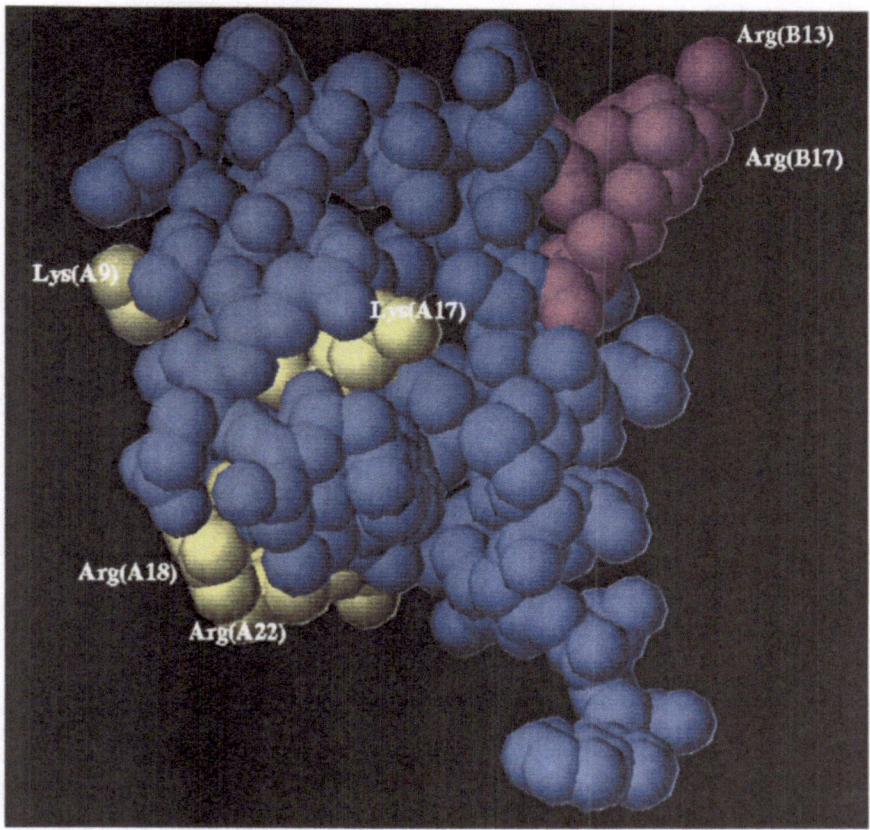

Fig. 11.5. X-ray structure of relaxin showing the location of the four basic residues of the A chain (yellow) in relation to the binding-site arginines on the B chain (magenta). The figure indicates that basic residues of the A chain are located on the surfaces opposite to the active site arginines.

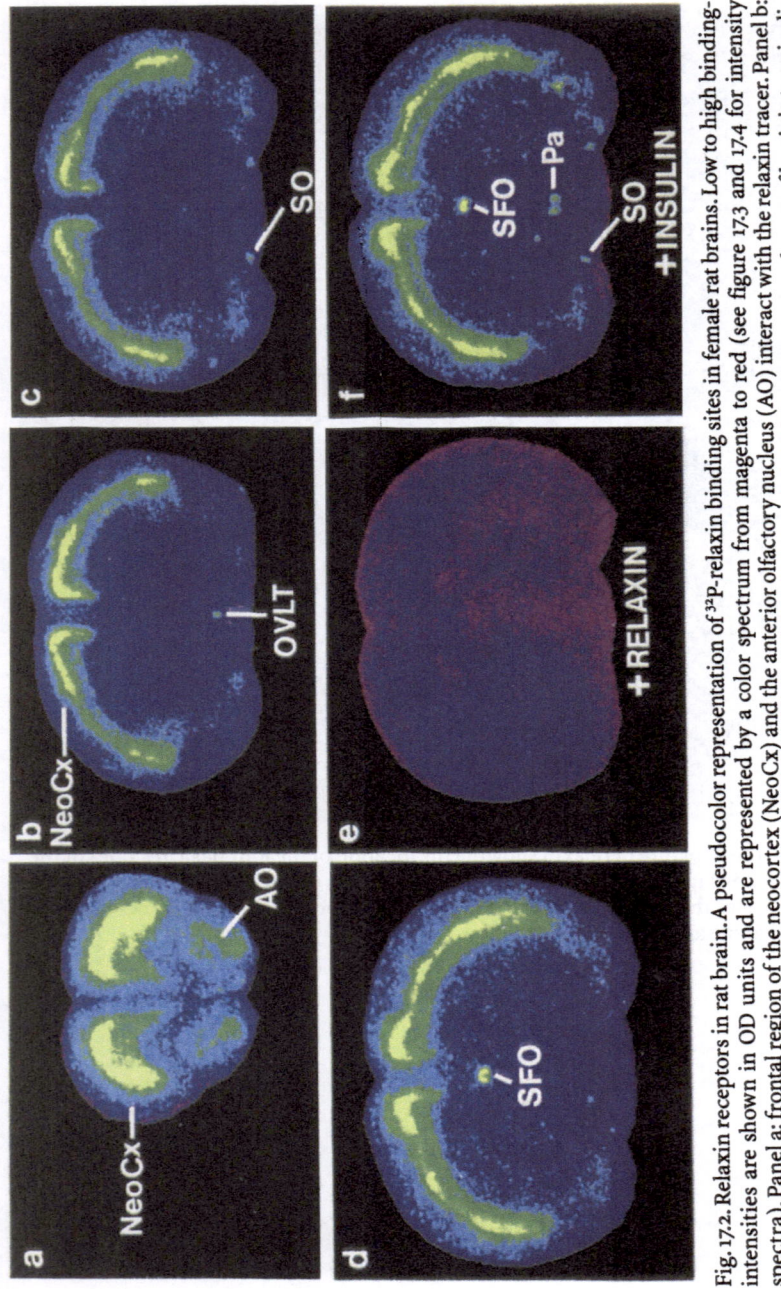

Fig. 17.2. Relaxin receptors in rat brain. A pseudocolor representation of ³²P-relaxin binding sites in female rat brains. Low to high binding-intensities are shown in OD units and are represented by a color spectrum from magenta to red (see figure 17.3 and 17.4 for intensity spectra). Panel a: frontal region of the neocortex (NeoCx) and the anterior olfactory nucleus (AO) interact with the relaxin tracer. Panel b: parietal region of the neocortex (NeoCx) shows relaxin-binding sites on the neocortex and the organum vasculosum of laminia terminalis (OVLT). Panel c: binding of ³²P-relaxin to the supraoptic nucleus (SO), panel d: binding to the subfornical organ (SFO), panel e shows the lack of binding in the presence of excess unlabeled relaxin. Panel f shows that radioactive relaxin binds to target tissue in the presence of insulin (Pa = paraventricular nucleus). Reprinted with permission from Osheroff PL, Phillips HS. Proc Natl Acad Sci USA 1991;88:6413–6417.

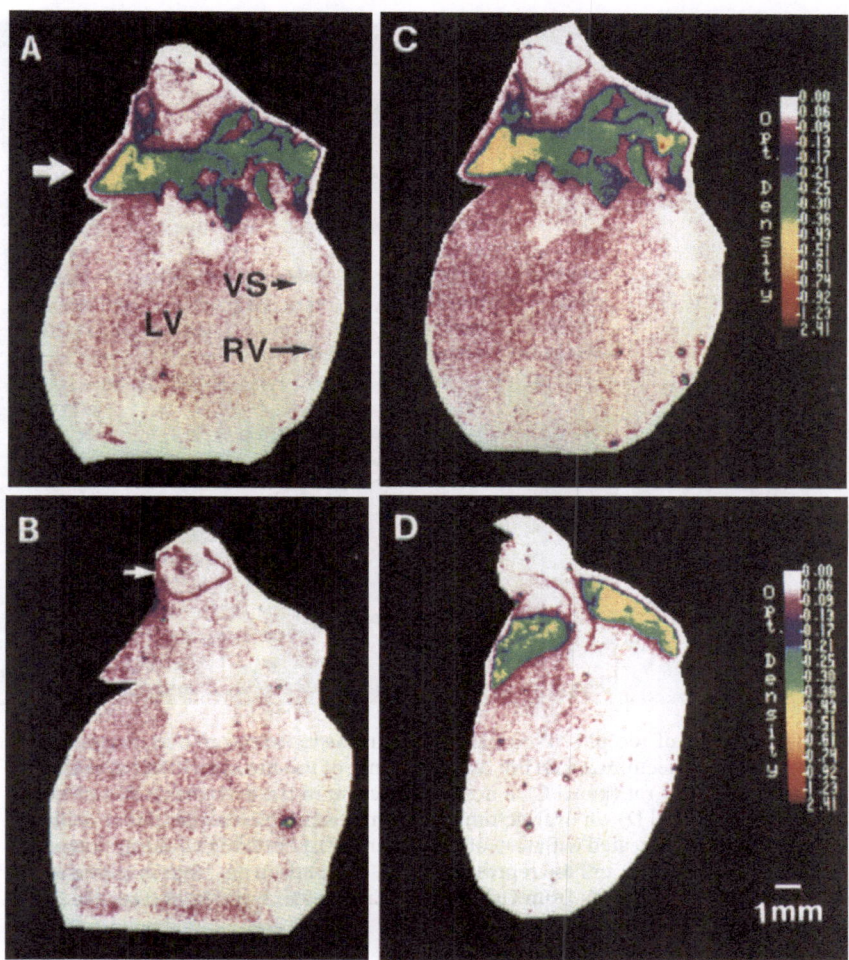

Fig. 17.3. Relaxin receptors in the rat heart. A pseudocolor representation of ^{32}P-relaxin-binding sites in female rat heart (panel A) in the presence of a 1000-fold excess of unlabeled relaxin (panel B) and 1000-fold excess of insulin-like growth factor I (panel C). Relaxi- binding to the male heart is shown in panel D. Tissue sections represent cross sections of the heart. (LV = left ventricals, RV = right ventricles, VS = ventricular septum). The arrow in panel A points to the atrium and the arrow in panel B points to the aorta. Low to high binding-intensities are shown in OD units and are represented by a color spectrum from magenta to red. Reprinted with permission from Osheroff PL, Cronin ML, Lofgren JA. Proc Natl Acad Sci USA 1992; 89:2384-2388.

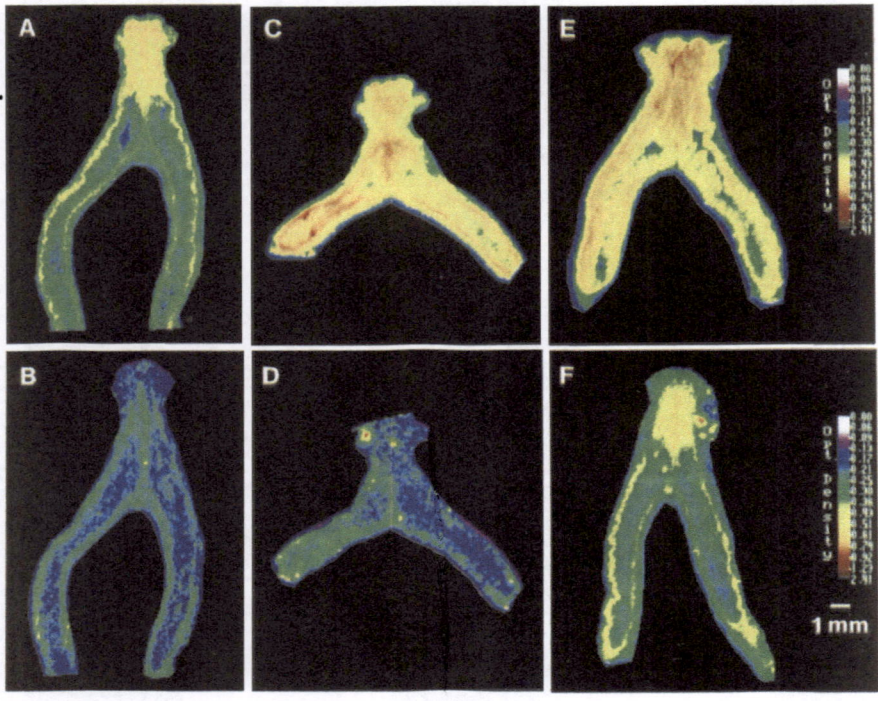

Fig. 17.4. Binding of 100 pM ^{32}P-human relaxin in the uterus of an ovariectomized rat (panel A), an ovariectomized rat but in the presence of 100 nM unlabeled relaxin (panel B), a normal intact rat (panel C), a normal intact rat in the presence of 100 nM unlabeled relaxin (panel D), an ovariectomized rat treated with estrogen (panel E) and an ovariectomized rat treated with testosterone (panel F). Low to high binding-intensities are shown in OD units and are represented by a color spectrum from magenta to red. Reprinted with permission from Osheroff PL, Cronin ML, Lofgren JA. Proc Natl Acad Sci USA 1992; 89:2384-2388.

Index